T0043467

Energy

A Beginner's Guide

ONEWORLD BEGINNER'S GUIDES combine an original, inventive, and engaging approach with expert analysis on subjects ranging from art and history to religion and politics, and everything in-between. Innovative and affordable, books in the series are perfect for anyone curious about the way the world works and the big ideas of our time.

Energy
A Beginner's Guide

Vaclav Smil

ONEWORLD

A Oneworld Paperback Original

First published by Oneworld Publications, 2006
This new revised edition published 2017
Reprinted 2017, 2018, 2019, 2020, 2021, 2022

Copyright © Vaclav Smil, 2006, 2017

The right of Vaclav Smil to be identified as the Author of this
work has been asserted by him in accordance with the Copyright,
Designs and Patents Act 1988

All rights reserved
Copyright under Berne Convention
A CIP record for this title is available from the British Library

ISBN 978-1-78607-133-0
eISBN 978-1-78607-134-7

Typeset by Silicon Chips
Printed and bound in Great Britain by Clays Ltd, Elcograf S.p.A.

Oneworld Publications
10 Bloomsbury Street
London, WC1B 3SR, England

Stay up to date with the latest books,
special offers, and exclusive content from
Oneworld with our newsletter

Sign up on our website
oneworld-publications.com

Energy will do anything that can be done in the world.
Johann Wolfgang von Goethe (1749–1832)

Contents

Contents

List of Figures

1

Energy in our minds: concepts and measures

The word energy is, as are so many abstract terms (from hypothesis to sophrosyne), a Greek compound. Aristotle (384–322 B.C.E.) created the term in his *Metaphysics*, by joining εν (in) and ἐργον (work) to form ἐνέργεια (*energeia*, "actuality, identified with movement") that he connected with *entelechia*, "complete reality." According to Aristotle, every object's existence is maintained by *energeia* related to the object's function. The verb *energein* thus came to signify motion, action, work, and change. No noteworthy intellectual breakthroughs refined these definitions for nearly two subsequent millennia, as even many founders of modern science had very faulty concepts of energy. Eventually, the term became practically indistinguishable from power and force. In 1748, David Hume (1711–1776) complained, in *An Enquiry Concerning Human Understanding*, that "There are no ideas, which occur in metaphysics, more obscure and uncertain, than those of *power, force, energy* or *necessary connexion*, of which it is every moment necessary for us to treat in all our disquisitions."

In 1807, in a lecture at the Royal Institution, Thomas Young (1773–1829) defined energy as the product of the mass of a body and the square of its velocity, thus offering an inaccurate formula (the mass should be halved) and restricting the term only to kinetic (mechanical) energy. Three decades later the seventh edition of the *Encyclopedia Britannica* (completed in 1842) offered only a very brief and unscientific entry, describing energy as "the power, virtue, or efficacy of a thing. It is also used figuratively, to denote emphasis in speech." Little has changed in popular discourse since that, or indeed since Hume's, time, except the frequency of the term's misuse. At the beginning of the twenty-first century energy, its derivative verb (energize) and its adjective (energetic), are used ubiquitously and loosely as qualifiers for any number of animated, zestful, vigorous actions and experiences, and energy is still routinely confused with power and force. Examples abound: a powerful new chairman brings fresh energy to an old company; a crowd is energized by a forceful speaker; pop-culture is America's soft power.

Devotees of physical fitness go one step further and claim (against all logic and scientific evidence) they are energized *after* a particularly demanding bout of protracted exercise. What they really want to say is that they feel better afterwards, and we have a perfectly understandable explanation for that: prolonged exercise promotes the release of endorphins (neurotransmitters that reduce the perception of pain and induce euphoria) in the brain and hence may produce a feeling of enhanced well-being. A long run may leave you tired, even exhausted, elated, even euphoric – but never energized, that is with a higher level of stored energy than before you began.

Science of energy: origins and abstracts

Sloppy use of ingrained terms is here to stay, but in informed writing there has been no excuse for ill-defined terms for more than a hundred years. Theoretical energy studies reached a satisfactory (though not perfect) coherence and clarity before the end of the nineteenth century when, after generations of hesitant progress, the great outburst of Western intellectual and inventive activity laid down the firm foundations of modern science and soon afterwards developed many of its more sophisticated concepts. The ground work for these advances began in the seventeenth century, and advanced considerably during the course of the eighteenth, when it was aided by the adoption both of Isaac Newton's (1642–1727) comprehensive view of physics and by engineering experiments, particularly those associated with James Watt's (1736–1819) improvements of steam engines (Figure 1; see also Figure 19).

During the early part of the nineteenth century a key contribution to the multifaceted origins of modern understanding of energy were the theoretical deductions of a young French engineer, Sadi Carnot (1796–1832), who set down the universal principles applicable to producing kinetic energy from heat and

Figure 1 James Watt

defined the maximum efficiency of an ideal (reversible) heat engine. Shortly afterwards, Justus von Liebig (1803–1873), one of the founders of modern chemistry and science-based agriculture, offered a basically correct interpretation of human and animal metabolism, by ascribing the generation of carbon dioxide and water to the oxidation of foods or feeds.

The formulation of one of the most fundamental laws of modern physics originates in a voyage to Java made in 1840 by a young German physician, Julius Robert Mayer (1814–1878), as ship's doctor. The blood of patients he bled there (the practice of bleeding as a cure for many ailments persisted well into the nineteenth century) appeared much brighter than the blood of patients in Germany.

Mayer had an explanation ready: blood in the tropics does not have to be as oxidized as blood in temperate regions, because less energy is needed for body metabolism in warm places. But this answer led him to another key question. If less heat is lost in the tropics due to radiation how about the heat lost as a result of physical work (that is, expenditure of mechanical energy) which clearly warms its surroundings, whether done in Europe or tropical Asia? Unless we put forward some mysterious origin, that heat, too, must come from the oxidation of blood – and hence heat and work must be equivalent and convertible at a fixed rate. And so began the formulation of the law of the conservation of energy. In 1842 Mayer published the first quantitative estimate of the equivalence, and three years later extended the idea of energy conservation to all natural phenomena, including electricity, light, and magnetism and gave details of his calculation based on an experiment with gas flow between two insulated cylinders.

The correct value for the equivalence of heat and mechanical energy was found by the English physicist (see Figure 2) James Prescott Joule (1818–1889), after he conducted a large number of careful experiments. Joule used very sensitive thermometers to measure the temperature of water being churned by an assembly of revolving vanes driven by descending weights: this arrangement made it possible to measure fairly accurately the mechanical energy invested in the churning process. In 1847 Joule's painstaking experiments yielded a result that turned out to be within less than one percent of the actual value. The law of conservation of energy – that energy can be neither created nor destroyed – is now commonly known as the first law of thermodynamics.

In 1850 the German theoretical physicist Rudolf Clausius (1822–1888) published his first paper on the mechanical theory of heat, in which he proved that the maximum performance

Figure 2 James Prescott Joule

obtainable from an engine using the Carnot cycle depends solely on the temperatures of the heat reservoirs, not on the nature of the working substance, and that there can never be a positive heat flow from a colder to a hotter body. Clausius continued to refine this fundamental idea and in his 1865 paper he coined the term *entropy* – from the Greek τρόπος (way) – to measure the degree of disorder in a closed system. Clausius also crisply formulated the second law of thermodynamics: entropy of the universe tends to maximum. In practical terms this means that in a closed system (one without any external supply of energy) the availability of useful energy can only decline. A lump of coal is a high-quality, highly ordered (low entropy) form of energy; its combustion will produce heat, a dispersed, low-quality, disordered (high entropy) form of energy. The sequence is irreversible: diffused heat (and emitted combustion gases) cannot be ever reconstituted as a lump of coal. Heat thus occupies a unique position in the hierarchy of energies: all other forms of energy can be completely converted to it, but its conversion into other forms can never be complete, as only a portion of the initial input ends up in the new form.

The second law of thermodynamics, the universal tendency toward heat death and disorder, became perhaps the grandest of all cosmic generalizations – yet also one of which most non-scientists remain ignorant. This reality was famously captured by C. P. Snow (1905–1980), an English physicist, politician, and novelist, in his 1959 Rede Lecture *The Two Cultures and the Scientific Revolution*:

A good many times I have been present at gatherings of people who, by the standards of the traditional culture, are thought highly educated and who have with considerable gusto been expressing their incredulity at the illiteracy of scientists. Once or twice I have been provoked and have asked the company how many

> of them could describe the Second Law of Thermodynamics.
> The response was cold: it was also negative. Yet I was asking
> something which is about the scientific equivalent of: "Have
> you read a work of Shakespeare's?"

Despite its supposed universality, the second law appears to be constantly violated by living organisms, whose conception and growth (as individuals) and whose evolution (as species and ecosystems) produces distinctly more ordered, more complex forms of life. But there is really no conflict: the second law applies only to closed systems under thermodynamic equilibrium. The Earth's biosphere is an open system, which incessantly imports solar energy and uses its photosynthetic conversion to new plant mass as the foundation for greater order and organization (a reduction of entropy).

Finally, the third law of thermodynamics, initially formulated in 1906 as Walther Nernst's (1864–1941) heat theorem, states that all processes come to a stop (and entropy shows no change) only when the temperature nears absolute zero (–273°C).

The first decade of the twentieth century brought a fundamental extension of the first law of thermodynamics when, in 1905, Albert Einstein (1879–1955) concluded that mass is itself a form of energy. According to perhaps the world's most famous equation – $E = mc^2$ – energy is equal to the product of mass and the square of the speed of light. As a result, just six tonnes of matter contain energy that is equivalent to the world's annual consumption of commercial energy in 2015 – but this astonishing potential remains just that, as we have no means to unlock the mass energy in limestone or water.

The only time when we commercially convert a relatively large (but still very small) share of mass into energy is in nuclear reactors: the fission (splitting) of the nuclei of one kilogram of uranium-235 releases an amount of energy equivalent to

190 tonnes of crude oil as it diminishes the initial mass by just one gram, or a thousandth of its original mass. In contrast, burning one kilogram of crude oil will diminish the mass of the fuel (and of the oxygen needed for its combustion) by only one ten billionth; too small a reduction to measure.

After less than a century of vigorous scientific effort the understanding of the nature of energy phenomena was virtually complete. But despite this large, and highly complex, body of scientific knowledge, there is no easy way to grasp the fundamental concept, which is intellectually much more elusive than is the understanding of mass or temperature. Richard Feynman (1918-1988), one of the most brilliant physicists of the twentieth century, put it with disarming honesty in his famous 1963 *Lectures on Physics*:

> It is important to realize that in physics today, we have no knowledge of what energy *is*. We do not have a picture that energy comes in little blobs of a definite amount. It is not that way. However, there are formulas for calculating some numerical quantity ... It is an abstract thing in that it does not tell us the mechanism or the *reasons* for the various formulas.

Difficult as it is, we have to try to make that abstract thing more accessible.

Fundamental concepts: energies, conversions, efficiencies

By far the most common definition of energy is "the capacity for doing work" but the full implication of this simple statement becomes clear only when you go beyond thinking about work as mechanical exertion– in physics terms, energy transferred through application of force over a distance, in common terms a labor to be performed, be it typing a letter or transplanting rice seedlings – and apply the term in a generic manner to any process that produces a change (of location, speed, temperature, composition) in an affected system (an organism, a machine, a planet). If you were to sit motionless in a quiet room for the next ten minutes, contemplating this statement, you would not have accomplished any work, in the narrow, strictly physical and commonly used, sense of applying force to a mechanical task.

But even as you sit motionless your metabolism is performing a great deal of work, as energy obtained from digested food is used (noting just the four key processes) to power your breathing, importing oxygen and exhaling carbon dioxide, to maintain the core temperature of your body at 37°C, to pump blood and to create the numerous enzymes that control everything from digestion to the transmission of nerve signals. By thinking hard about an abstract concept you do actually use a bit more energy but making all those additional neuronal connections in your brain amounts to an entirely negligible mark-up. Even when you are fast asleep, your brain accounts for about twenty percent of your body's metabolism and even a taxing mental exercise will raise that share only a little.

Outside a quiet room, the work done by various energies is accomplished in myriad ways. The lightning that slashes through summer skies works in a very different way from a giant harbor crane picking up large steel crates from a pier and stacking them up to a dizzying height on a container ship – and the differences are due to one of the most fundamental physical realities, the existences of multiple forms of energy and their conversions, on space and time scales ranging from galactic to sub-atomic and from evolutionary to ephemeral. Lightning works in a tiny fraction of a second, illuminating and heating the atmosphere and decomposing molecules of nitrogen, that is, converting the electrical energy of cloud-to-cloud or cloud-to-earth discharge to electromagnetic, thermal, and chemical energy. In contrast, the motors of stacking cranes in container ports work around the clock, converting electrical energy into mechanical energy and then into the potential energy of loaded cargo.

Energy is not a single, easily definable entity, but rather an abstract collective concept, adopted by nineteenth-century physicists to cover a variety of natural and anthropogenic (generated by humans) phenomena. Its most commonly encountered forms are heat (thermal energy), motion (kinetic or mechanical energy), light (electromagnetic energy) and the chemical energy of fuels and foodstuffs. Some of their conversions are the very fundamentals of life: during photosynthesis a small share of the electromagnetic energy of light becomes the chemical energy of bacteria and plants, and cooking and heating is done by converting chemical energy in biomass (wood, charcoal, straw) or fossil fuels (coals, oils, gases) to thermal energy (Figure 3). Others are a matter of convenience enjoyed on large scales: the conversion of chemical energy to electrical energy in batteries operates billions of mobile phones, music players, and radios. And some are quite rare: for example, the gamma-neutron reactions that are produced by converting electromagnetic energy to nuclear energy are used only for specialized scientific and industrial tasks.

Kinetic energy is associated with all moving masses, be they heavy, tank-penetrating shells made of depleted uranium or wispy clouds ascending above a tropical rainforest. Its manifestations are easy to perceive and its magnitude easy to calculate, as it is simply half of the moving object's mass (m) multiplied by the square of its velocity (v): $E_k = \frac{1}{2}\ mv^2$. A key thing to note is that kinetic energy depends on the square of the object's velocity: doubling the speed imparts four times more energy, tripling it nine times more – and hence at high speed, even small objects can become very dangerous. Tornado winds, in excess of 80 meters per second (nearly 290km/h) can drive featherweight pieces of straw into tree trunks; tiny space debris (a lost bolt) traveling at 8,000m/s could pierce the pressurized suit of a space-walking

from → to ↓	electro-magnetic	chemical	thermal	kinetic	electrical	nuclear	gravitational
electro-magnetic		chemilumines-cence	thermal radiation	accelerating charge phosphor	electro-magnetic radiation electro-luminescence	gamma reactions nuclear bombs	
chemical	photo-synthesis photo-chemistry	chemical processing	boiling dissociation	dissociation by radiolysis	electrolysis	radiation catalysis ionization	
thermal	solar absorption	combustion	heat exchange	friction	resistance heating	fission fusion	
kinetic	radiometers	metabolism muscles	thermal expansion internal combustion	gears	motor electro-strictions	radioactivity nuclear bombs	falling objects
electrical	solar cells photo-electricity	fuel cell batteries	thermo-electricity thermionics	conventional generator		nuclear batteries	
nuclear	gama-neutron reactions						
gravitational				rising objects			

Figure 3 Energies and their conversions

astronaut, and (although the risk has turned out to be very small indeed) a space vehicle can be damaged by a micrometeoroid traveling at 60,000 m/s.

Potential energy results from a change in the spatial setting of a mass or of its configuration. Gravitational potential energy, resulting from a changed position in the Earth's gravitational field, is ubiquitous: anything lifted up acquires it, be it rising water vapor, a hand lifted in a gesture, a soaring bird, or an ascending rocket. Water stored behind a dam in order to fall on turbine blades and generate electricity is a practical example of using gravitational potential energy to a great economic advantage:

nearly twenty percent of the world's electricity is generated this way. The potential energy of water stored behind a dam (or a rock precariously poised on a weathering slope) is simply a product of the elevated mass, its mean height above ground (h) and the gravitational constant (g): E_p = mgh. Springs that have been tensioned by winding are a common example of the practical use of elastic potential energy that is stored due to their deformation and released as useful work as the coil unwinds and powers a watch or a walking toy.

Biomass (living, in plants, micro-organisms, animals, and people; and dead, mainly in soil, organic matter, and tree trunks) and fossil fuels (formed by the transformation of dead biomass) are enormous stores of chemical energy. This energy is held in the atomic bonds of tissues and fuels and released through combustion (rapid oxidation) which produces heat (an *exothermic* reaction). This results in new chemical bonds, the formation of carbon dioxide, frequently the emission of nitrogen and often sulfur oxides, and, in the case of liquid and gaseous fuels, the production of water.

HEAT

Heat of combustion (or specific energy) is the difference between the energy of the bonds in the initial reactants and that in the bonds in the newly formed compounds. The poorest fuels (wet peat, wet straw) release less than a third of the thermal energy produced by burning gasoline or kerosene. The energy content of a fuel, foodstuff, or any other combustible material can easily be determined by burning an absolutely dry sample of it in a calorimeter (a device that measures the heat given off during chemical reactions). Heat is produced by a number of other energy conversions: nuclear fission is a major commercial process whose heat is used to generate electricity, heat arising due to the resistance to the flow of electric current is used to prepare food, boil water, and warm interior spaces, and friction produces a great deal of unwanted (inside vehicle transmissions) as well as unavoidable (between vehicle tires and road) heat.

HEAT (cont.)

Once produced, heat can be transferred in three ways: conduction (that is direct molecular contact, most commonly in solids), convection (by moving liquids or gases) and radiation (the emission of electromagnetic waves by bodies warmer than their surrounding). Most of the heat that radiates at ambient temperatures from the Earth's surface, plants, buildings, and people is invisible infra-red radiation, but hot (above 1,200°C) objects (such as the coiled tungsten wires in light bulbs, molten steel in electric arc furnaces or distant stars) radiate invisible light, also.

Latent heat is the amount of energy needed to effect a physical change with no temperature change: changing water to steam (latent heat of vaporization) at 100°C requires exactly 6.75 times more energy than the changing of ice into water at 0°C.

The heating of water also accounts for most of the difference between the gross (or higher) heating value of fuels and their net (lower) heating value. The first is the total amount of energy released by a unit of fuel during combustion with all the water condensed to liquid (and hence the heat of vaporization is recovered); the second subtracts the energy required to evaporate the water formed during combustion. The difference is around one percent for coke (virtually pure carbon, whose combustion generates only carbon dioxide), around ten percent for natural gases and nearly twenty percent for pure hydrogen (whose combustion generates only water). The gap may be even larger for wood, but only a small part of the difference is due to hydrogen present in the fuel. Fresh (wet) wood simply contains too much moisture (sometimes more than seventy-five percent): most of the thermal energy released by the combustion of unseasoned (green) wood goes into evaporating water rather than warming a room and if wet wood has more than sixty-seven percent of water it will not ignite.

The efficiency of energy conversion is simply the ratio of desirable output to initial input. Photosynthesis is perhaps the best example of a highly inefficient process: even for the most productive crops no more than four to five percent of the solar radiation that strikes their fields every year will be transformed into new plant mass (phytomass), and the global annual average of the process (commonly prevented by excessive cold or lack of moisture) equates to a meager 0.3 percent. When the initial input is limited only to photosynthetically active radiation (wavelengths that can be absorbed by plant pigments, which average about forty-five percent of the incoming sunlight) the useful transfer doubles but globally still remains below one percent. High energy loss during a low-efficiency conversion simply means that only a very small part of the original energy input could be transformed into a desired service or product: no energy has been lost (the first law of thermodynamics), but (as the second law of thermodynamics dictates) a large share of the initial input ends up as unusable, dispersed heat.

In contrast, there is no shortage of processes, devices, and machines whose efficiency is greater than ninety percent. Electricity can be converted to heat by a baseboard resistance heater with one hundred percent efficiency. Healthy people on balanced diets can digest carbohydrates (sugars, starches) with an efficiency of as much as ninety-nine percent, the best natural gas furnaces can convert between ninety-five and ninety-seven percent of the incoming fuel into heat inside a house, more than ninety-five percent of electricity gets converted into the rapid rotation of large electrical motors, and, conversely, the giant turbines in thermal stations convert up to ninety-nine percent of their mechanical energy into electricity as they rotate in a magnetic field.

Despite their diverse manifestations − ranging from the blinding light of our nearest star to the imperceptible but deadly ionizing radiation that can escape from a malfunctioning nuclear reactor, from the high-temperature combustion in rocket engines to the amazingly intricate enzymatic reactions that proceed at ambient temperature and pressure − all energy phenomena can be quantified with the help of a small number of universal units. While many traditional yardsticks are still in everyday use around the world, modern scientific and engineering quantifications are based on the *Système*

International d'Unités (International System of Units, commonly abbreviated as SI) that was adopted in 1960. In this book I will use only the appropriate SI units: the complete list, as well as the prefixes to indicate multiples and submultiples, will be found later in this chapter.

Quantitative understanding: the necessity of units

The SI specifies seven fundamental measures: length, mass, time, electric current, temperature, amount of substance, and luminous intensity. These units are used directly, to measure the seven common variables, as well as to derive more complex quantities. The latter category includes some relatively simple units used in everyday situations (area, volume, density, speed, pressure) as well as more complex concepts deployed in science and engineering (force, pressure, energy, capacitance, luminous flux). Only three fundamental variables – mass (M), length (L), and time (T) – are needed to derive the units repeatedly encountered in energy studies. Area is obviously L^2, and volume L^3, mass density M/L^3, speed L/T, acceleration (change of speed per unit of time) L/T^2, and force, according to Newton's second law of motion, ML/T^2 (mass multiplied by acceleration). Energy is expended (work is done) when a force is exerted over a distance: energy's dimensional formula is thus ML^2/T^2.

The scientific definition of *power* is simply the rate of energy use: power equals energy per time, or ML^2/T^3. One of the most common abuses of the term, found even in engineering journals, is to confuse power with electricity and to talk about power generating plants: in reality, they generate electrical energy at a variable rate, determined by industrial, commercial, and household demand for kinetic energy (produced by electric motors), thermal energy (for industrial furnaces, heat processing, and home heating) and electromagnetic energy (or more accurately its visible segment, light). And, obviously, knowing a particular machine's power rating tells you nothing about how much energy it will use unless you know for how long it will run.

Everybody is familiar with the standard names of SI units for length (meter, m), mass (kilogram, kg) and time (second, s) but degrees Kelvin (K) rather than Celsius are used to measure temperature; the ampere (A) is the unit of electric current, the mole (mol) quantifies the amount of substance and the candela (cd) the luminous intensity. More than twenty derived units, including all energy-related variables, have special names and symbols, many given in honor of leading scientists and engineers. The unit of force, kgm/s^2 (kilogram-meter per second squared), is the newton (N): the application of $1\,N$ can accelerate a mass of one kilogram by one meter per second each second. The unit of energy, the joule (J), is the force of one newton acting over a distance of one meter (kgm^2/s^2). Power, simply the energy flow per unit of time (kgm^2/s^3), is measured in watts (W): one watt equals one J/s and, conversely, energy then equals power 3 times, and hence one J is one watt-second.

Unit of	Name	Symbol
Length	meter	m
Mass	kilogram	kg
Time	second	s
Electric current	ampere	A
Temperature	kelvin	K
Amount of substance	mole	mol
Luminous intensity	candela	cd

BASIC SI UNITS

One of the most revealing measures in energy studies is power density (W/m^2). This SI-derived unit is sometimes called, in a more restrictive manner, heat flux density or irradiance, but the concept of power per unit area can obviously be extended to any flow of energy, from food harvests to average demand for electricity in densely inhabited urban areas. The measure's denominator is the area of the Earth's surface, a building's footprint, or any other horizontal area. The power density of incoming solar radiation determines the biosphere's energy flows; the power density of household energy use dictates the rate of fuel and electricity inputs. In some cases it is also revealing to calculate the vertical power density of energy flows. This is particularly useful in the case of the strong winds, floods, and tsunami that can strike vegetation and structures with large forces per unit of vertical area and cause tremendous damage: just think of the March 11, 2011 tsunami generated by the Tōhoku earthquake.

Perhaps the easiest way to get an appreciation for the magnitude of these energy and power units is through gravitational acceleration: at the Earth's surface this equals $9.81\,m/s^2$; rounding this to ten (boosting it by less than two percent) will make the following calculations easier. If you hold a mass of one kilogram one meter above ground – for example a plastic one-liter bottle of water roughly at the elbow height – it will be subject to a force (downward gravitational pull) of ten newtons. If you hold instead something that has only one-tenth of the bottle's mass (a small mandarin orange is about $0.1\,kg$) it will be subject to the force of one newton. So, picking up that orange from the kitchen floor and putting it on the counter (roughly one meter above the floor) requires the expenditure of one joule of energy; if you did it in about one second then you would have expended energy at the rate of one watt.

This is an apposite place to reiterate that the power tells you nothing about the total amount of energy consumed or released by the rated process. A giant lightning bolt has a power of the same order of magnitude (10^{13}W) as the Earth's geothermal flow – but the lightning is ephemeral, lasting a very small fraction of a second, while the geothermal flow has been going on incessantly since the planet's formation some four and a half billion years ago. Similarly, if you are a small (50 kg) female, your basal metabolism (the conversion of food into energy forms that can be used for growth and activity), going non-stop as long as you live, amounts to some 60 W – a rate as small as a lamp that you may switch on only occasionally for a few hours. The solar radiation reaching the Earth is, of course, its most powerful continuous energy flow, which delimits most natural processes (geothermal energy and gravitational attraction do the rest) and hence the characteristics of the planet's habitable environment: it proceeds at the rate of 1.73×10^{17} W (that is 173 PW). In contrast, in 2015, the consumption of all fossil fuels added up to a global rate of 15 TW, the equivalent of only about 0.009 percent of the planet's received solar radiation.

MAGNITUDES OF ENERGY AND POWER

Basic energy and power units refer to very small amounts and rates. A single chickpea contains 5,000 J of chemical energy; a tiny vole needs 50,000 J a day just to stay alive. The full gasoline tank in my Honda Civic contains about 1,250,000,000 J and when I drive I burn that fuel at the rate of about eight liters per 100 km, which equates to an average power of about 40,000 W. Winds in a violent thunderstorm will unleash more than 100,000,000,000,000 J in less than an hour so their power will be more than 25,000,000,000 W. The need for specific prefixes to avoid writing out all those zeros or using constantly scientific notation ($10 n$) is thus obvious and, given the smallness of basic units, energy studies uses not only the common thousand – (10^3, kilo, k) and million-fold (10^6, mega, M) prefixes but also higher multipliers: G (10^9, giga), T (10^{12}, tera), P (10^{15}, peta), and E (10^{18}, exa). New prefixes, for 10^{21} and 10^{24}, were added to the SI in 1991.

Mega, giga (MJ and GJ), and kilo (kWh) are the most commonly used multipliers for energy, kilo, mega, and giga

MAGNITUDES OF ENERGY AND POWER (*cont.*)

Multiples

Prefix	Abbreviation	Scientific notation
deka	da	10^1
hecto	h	10^2
kilo	k	10^3
mega	M	10^6
giga	G	10^9
tera	T	10^{12}
peta	P	10^{15}
exa	E	10^{18}
zeta	Z	10^{21}
yota	Y	10^{24}

(kW, MW, and GW) for power. The net energy content of fuels ranges from eleven MJ/kg (or GJ/t) for air-dry straw (about twenty percent moisture) to forty-four MJ/kg for gasoline, and the gross energy content of foods (digestibility determines the share of food that is actually used by an organism) goes from less than one MJ/kg for leafy vegetables to nearly forty MJ/kg for pure fats (a table later in this chapter lists the averages and ranges of energy contents for all common fuels and major foodstuff categories). One thousand watt-hours or 3.6 million watt-seconds are one kilo-watthour (kWh), a unit commonly used to measure and price electricity consumption: the average American household uses about 1,000 kWh (1 MWh) a month, roughly the equivalent of having fourteen 100 W lights ablaze night and day for thirty days.

As for power, small kitchen appliances (from coffee grinders to coffee makers) are mostly rated between 50–500 W, the power of passenger cars has been increasing and in 2015 it was between 90 kW for subcompacts (Ford Fiesta) and 120 kW for compacts

MAGNITUDES OF ENERGY AND POWER (*cont.*)

Energy content of fuels	MJ/kg
Hydrogen	114.0
Gasolines	44.0–45.0
Crude oils	42.0–44.0
Natural gas	33.0–37.0
Anthracite	29.0–31.0
Bituminous coal	22.0–26.0
Lignites	12.0–20.0
Air-dried wood	14.0–16.0
Cereal straws	12.0–15.0

Submultiples

Prefix	Abbreviation	Scientific notation
deci	d	10^{-1}
centi	c	10^{-2}
milli	m	10^{-3}
micro	μ	10^{-6}
nano	n	10^{-9}
pico	p	10^{-12}
femto	f	10^{-15}
atto	a	10^{-18}
zepto	z	10^{-21}
yocto	y	10^{-24}

MAGNITUDES OF ENERGY AND POWER (*cont.*)

(Honda Civic), while Toyota Camry, a popular sedan, rates up to 200 kW and the world's most powerful car, the limited production plug-in hybrid Koenigsegg Regera reaches an obscene 1.1 MW! Large steam- and water-driven turbo generators have capacities of between 500 and 1,650 MW and their multiple installations in the world's largest fossil-fueled power plants can generate electricity at rates surpassing 2 GW. China's Sanxia (Three Gorges) project (the world's largest) has thirty-four turbines with an aggregate capacity of 22.5 GW.

Common power density yardsticks include the total amount of solar radiation reaching the ground (averaging about 170 W/m²) and the thermal energy radiated by downtowns of large cities (the urban heat island effect, commonly in excess of 50 W/m²). As far as vertical power densities are concerned, well-built structures should not suffer any notable damage from fluxes below 18 kW/m²; powerful tornadoes strike with more than 100 W/m² and tsunami can be even more destructive.

At the opposite end of the power and energy spectrum are the quantities that need the most commonly used submultiples: milli (m, 10^{-3}), micro (μ, 10^{-6}), and nano (n, 10^{-9}). Every strike as I typed this book cost me about 2 mJ of kinetic energy, a 2 mm dewdrop on a blade of grass has a potential energy of 4 μJ, and the mass-energy of a proton is 0.15 nJ. Power-wise, the laser in a CD-ROM drive is rated at 5 mW, a quartz watch needs about 1 μW, and a flea hops with the power of some 100 nW.

NON-SI UNITS

All standard SI units have traditional (imperial) counterparts, still used by many craftsmen and engineers in English-speaking countries. The energy content of fuels is still commonly expressed in British thermal units (one Btu = 1055 J), work accomplished in foot-pounds-force (one ft-lbf = 1.36 J), power (of cars or motors) in horsepower (one hp = 745 W), and force in pounds (one lb force = 4.45 N).

NON-SI UNITS (*cont.*)

There is also one metric but non-SI unit not derived from the seven basic measures: the calorie is the amount of heat needed to raise the temperature of 1g of water from 14.5 to 15.5°C. This is a small unit of energy, equal to just over four J (1 cal = 4.18J), and so we most often use its 1,000-fold multiple, a kilocalorie (kcal). A healthy, active adult male with a body mass index (calculated by dividing body weight in kg by the square of height in m) within the optimum range (19–25) will need about 2,500 kcal (2.5 Mcal or 10.5 MJ) of food a day.

But, instead of using the proper scientific prefix, nutritionists began to use Cal (large calorie) to signify a kilocalorie; because small c has been often used mistakenly instead of the capital letter, people are unaware of the difference. You may have friends arguing with you that all you need to eat is 2,500 calories a day. Set them straight: that amount would not feed a twenty gram mouse. Even its daily basal metabolism (assuming it could lie motionless for twenty-four hours – not an easy feat for a mouse) requires about 3,800 cal (almost 16 kJ) a day. In contrast, the daily basic metabolic rate of a healthy 70 kg adult male is about 7.1 MJ and activities will increase this total by anywhere between twenty (for a sedentary lifestyle) and one hundred percent for prolonged heavy exertion, such as demanding physical labor or endurance sports.

Turning to electricity, current (the flow of electrons through a conductor, usually labeled I in equations) is measured in amps (A): André-Marie Ampère (1775–1836), a French mathematician, was one of the founders of modern electrodynamic studies. The volt (V) – named after Alessandro Volta (1745–1827), an early experimenter with electricity and inventor of the first battery – is the derived unit ($V = W/A$) of electrical potential, and thus a measure of the difference between the positive and negative terminals of a battery. The resistance (R) encountered by a current is measured in ohms (Ω) and depends on the conducting material and its dimensions. Copper is a nearly seventy percent better conductor than pure aluminum which, in turn, conducts just over three times better than pure iron, and long thin wires offer more resistance than short thick ones. But aluminum alloys are much cheaper than pure copper and so we use them, rather than copper, for long-distance high-voltage lines.

In direct current (DC), electrons flow only in one direction, while alternating current (AC) constantly changes its amplitude and reverses its direction at regular intervals: in North America it does so 120 times a second (a frequency of 60 cycles per second), in Europe, 100 times a second. Ohm's law – Georg Simon Ohm (1789–1854) was a German mathematician and physicist – relates voltage to resistance and current in DC circuits in a linear way: $V = IR$. The law has to be modified for AC circuits because it ignores reactance, the opposition encountered by the flow of AC in coils (inductive reactance) and capacitors (capacitive reactance). Using impedance (Z, the combined opposition of reactance and resistance, also measured in Ω), the modified law becomes $I = V/Z$. But using unadjusted, Ohm's law will not make any major difference for such common household electricity converters as lights and appliances.

This relationship has profound implications both for transmitting electricity and for using it safely. Electric shocks, and the risk of electrocution, depend on the current that passes through a body, above all. According to Ohm's law, $I = V/R$, which means that for any given electricity voltage (household supply is 120V in North America, 230V in most of Europe) the current will be minimized by higher resistance. Dry skin offers a resistance of more than $500\,k\Omega$ and will limit the current to a harmless level of just a few milliamps. In contrast, wet skin provides a low-resistance (just $1\,k\Omega$) conductor for lethal currents of 100–300 mA, which can trigger ventricular fibrillation and death. Accidentally touching a 120V wire with dry hands is rarely lethal for healthy adults; poking about with sweaty hands and bare feet on a humid summer day is an altogether different proposition, even with just 120V. And, all else being equal, one is always better off touching a live wire in Boston or Vancouver than in London or Paris.

DIRECT AND ALTERNATING CURRENT

Because power is the product of current (amperes) and voltage (volts), and voltage equals current multiplied by resistance (ohms), power is the product of the current squared and resistance (I^2R) – and current and resistance together determine your use of electricity. For example, you need a high resistance (about 140 ohms) to produce a white glow of incandescent light in vacuum, and a relatively low resistance (about 15 ohms) to heat wires in a bread toaster. But a light bulb needs only 100 W and so it draws only about 0.8A; in contrast, the toaster rates 800W and so it needs more than 7A. This also means that if an electricity-generating plant of a given capacity is to transmit this power through wires of a given resistance then increasing the voltage will reduce the transmission losses in proportion to the square of V.

This explains why all modern networks use AC both for long-distance transmission and distribution to homes. The earliest electric networks, engineered by Thomas Edison (1847–1931) during the early 1880s, delivered DC, whose voltage either had to match that of the load (a light or a motor) or be reduced to its level by a converter placed in series or a resistor that wasted the difference. Raising voltage and reducing current, in order to limit DC transmission losses, would have resulted in dangerously high load voltages in houses and factories. In contrast, even after transmitting high-voltage AC across long distances with minimized losses, it can be reduced to acceptably low voltages by transformers.

Edison resisted the introduction of AC until 1890 and indeed actively campaigned against it. The innovations of the late 1880s – reliable transformers, AC motors and meters and rotary DC-to-AC converters that made it possible to connect existing DC stations and networks to high-voltage AC lines – decided the outcome: the battle of systems was basically over by 1890, and although some DC networks lingered until after World War I, the future clearly belonged to AC. But there is plenty of DC coursing through electrical devices all around us, either converted from AC or supplied by batteries. High starting torque (a force that produces rotation) makes DC motors the best choice for electric trains. They get their supply from overhead AC lines and convert it, onboard, to DC, and personal computers use converters to supply up to 5 V DC to the digital circuits and more than 10V to the motors for disk drives.

DIRECT AND ALTERNATING CURRENT (*cont.*)

Portable DC needs are supplied by many types of batteries, compact devices that change chemical energy directly into electrical energy. By far the most common is the big, rechargeable, lead-acid battery used in hundreds of millions of cars: it supplies 12V from six cells, with cathodes (positive electrodes) of lead and anodes (negative electrodes) of lead oxide-coated lead and between them, sulfuric acid. Car batteries energize not only starter motors but also the many small DC motors that do tasks that were previously done by hand (open windows, adjust side-view mirrors, lock doors). The small cylindrical batteries that power an increasing array of toys, flashlights, radios, televisions, and music players are fundamentally just improved versions of the carbon-zinc cell invented by Georges Leclanché (1839–1882) in the 1860s. His original battery had a wet electrolyte, the familiar dry cell that uses a slightly acidic pasty electrolyte came later. In 1959, Energizer introduced an alkaline battery (with potassium hydroxide electrolyte) and replaced the traditional carbon (graphite) cathode with manganese dioxide. All cylindrical batteries have a flat metal base as a negative and a raised metal cap as a positive terminal. The most common types are the chunky D for flashlights and finger-slim AA for electronic gadgets. All these small cylindrical batteries deliver 1.5V at 2,600mA/h, and have a fairly long shelf life, but their voltage steadily drops with use. There are many other types of batteries including slim lithium ion rectangular prisms for laptop computers (from 6–16V) and the tiny silver oxide 1.5V button cells that power hearing aids and watches.

Now we are ready to proceed, in evolutionary sequence, with a systematic review of energies in nature, history, and modern society.

2

Energy in the biosphere: how nature works

There would be no life on the Earth without the sun but a planet orbiting a star is actually extremely unlikely to have a biosphere, a thin sliver of space that harbors life and allows its evolution. The probes we sent to Mars did not uncover any evidence of life; Venus, our other neighbor, is too hot, and the remaining planets of the solar system are even less suitable for sustaining the only life we know of: carbon-based organisms that encode, in nucleic acids, complex programs for their reproduction and survival and run their metabolism with the aid of enzymes. Although we have been discovering many new extra-solar planets (planets orbiting another star) we have no indication that any of them support life (most are simply too big), and despite considerable resources invested in listening to space sounds we have not heard from anybody; all we hear are the radio waves emitted by the ionized interstellar gas that surrounds hot stars.

None of this is too surprising, considered in strictly energetic terms. What is needed is not simply a star (our galaxy alone has some hundred billion) with orbiting planets (again, they must be quite numerous) but a suitable "Goldilocks" star: not too big, not too small, not too cold, not too hot. Stars that are too massive do not last long enough to allow for the billions of years that, in the Earth's case, were needed to produce complex life, and long-lived, dwarfish stars have insufficient luminosity to energize any planets that may orbit them. A Goldilocks star then has to capture a Goldilocks planet; one that is not too far away to have its water frozen all the time (like Mars), nor too close to have it vaporized (like Venus).

And that is merely the beginning of a long list of prerequisites that must be satisfied to achieve the conditions that make a planet habitable, or even the simplest life possible. The best way to demonstrate this is to imagine changing just one of the attributes that influence the delivery of solar energy to the Earth (playing the "everything-else-being-equal" game). What if gravity were twice as strong? What if its orbit were far more eccentric than its actual, nearly circular, course? What if its axis of rotation were not inclined? What if its rotation took 240 days instead of twenty-four hours? What if ninety (rather than thirty) percent of its surface were land? What if it had no water vapor

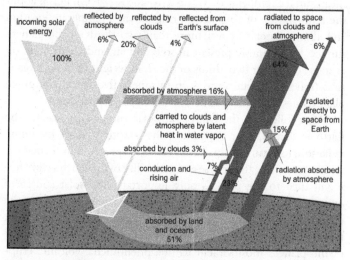

Figure 4 The Earth's radiation balance (based on NASA image)

and no carbon dioxide in its atmosphere? I will let you speculate about the unwelcome consequences of the first five ifs and give you just the answer for the last one: there would be no life on the Earth.

Our planet's atmosphere lets incoming radiation (except for the shortest wavelengths) reach the surfaces and warm them (see Figure 4) but absorbs, temporarily, part of the outgoing longwave radiation. Without this absorption the Earth would be a "perfect black body radiator" and would re-radiate all the intercepted solar energy, leaving the planet at a temperature of 255 K (−18°C). At −18°C water would be permanently frozen, and there could be no life. Atmospheric gases, which selectively absorb part of the outgoing radiation (and then re-radiate it down and up) change a planet's average radiative temperature (this is what makes Venus too hot and Mars too cold). On Earth, this "greenhouse effect", caused mostly by water vapor and CO_2, has been just right for the evolution and diversification of complex life, as it raises the mean surface temperature, by 33 K, to 288 K (15°C). This is about the temperature of a pleasant spring day in a temperate location and allows water to cover more than two-thirds of the planet's surface, be copiously present in soils and air, and to account, on average, for about two-thirds of the fresh weight of living organisms (much more in some − up to ninety-five percent in green plant tissues and ninety-nine percent in phytoplankton).

After following the fate of the solar energy that reaches the Earth − that is, after examining the partitioning of incoming radiation, its return to space, and its atmospheric transformation, which determine the course and variability of the Earth's climate − I will turn to the planet's only important non-solar source of energy, its internal heat. This geothermal energy powers the global tectonics whose manifestations include not only a constant refashioning of the Earth's continents and oceans but also some of the planet's most violent natural phenomena: volcanic eruptions, earthquakes, and tsunami.

Despite stunning external biodiversity, living organisms share a relatively small number of fundamental metabolic pathways that use available energy to convert simple inputs into new living mass. *Autotrophs* (also called primary producers, which include all organisms able to synthesize new biomass from simple inorganic compounds) use two distinct ways to produce new biomass. *Phototrophs* (terrestrial plants, algae, phytoplankton, cyanobacteria, and green and purple sulfur bacteria) convert electromagnetic energy into high-energy phosphate bonds within ATP (adenosine triphosphate, the molecule chiefly responsible for the store and transport of energy within cells) and use this energy to produce new mass (phytomass) from atmospheric CO_2 and the macronutrients (nitrogen, phosphorus, and potassium) and micronutrients (iron, calcium, silicon, and others) available in the soil. *Chemotrophs* (nitrifying bacteria (the bacteria which turn ammonia into nitrates), iron bacteria, non-pigmented sulfur bacteria, and methane-producing microbes) do not need light, only CO_2, oxygen, and either an oxidizable element (hydrogen, iron) or a simple inorganic compound (hydrogen sulfide, ammonia). Their invisible metabolism is indispensable for the biosphere's critical biological, geological, and chemical (biogeochemical) cycles.

Heterotrophs (also called chemo-organotrophs) are organisms that cannot synthesize new biomass from simple inorganic inputs and must get their building blocks from digested organic compounds: this includes most bacteria, fungi, and animals. Heterotrophs fall into four principal categories: primary consumers (herbivores), secondary and higher level consumers (carnivores), consumers of dead and decaying biomass (saprovores, detritivores), and species that resort to all the above strategies (omnivores). Modern energy studies have uncovered both great commonalities in plant and animal metabolism and many fascinating niche adaptations, and have also traced complex energy flows both on the large scale of individual ecosystems and the global scale of grand biogeochemical cycling, particularly the carbon and nitrogen cycles.

Sun and earth: solar radiation and its conversions

Astronomers like to point out that the sun belongs to one of the most common types of star (G2 dwarfs) unremarkable either for their size or their radiation. Most of their power comes from the proton-proton reaction, the fusion of hydrogen atoms into helium that proceeds at temperatures greater than 13 million K. The sun's energy production (total luminosity) is immense, seen in terrestrial terms, as thermonuclear reactions in its core convert some 4.4 Mt of matter into energy every second: according to Einstein's mass-energy equation, this works out to nearly 3.9×10^{26} W, a rate thirteen orders of magnitude (roughly 30 trillion times) greater than our use of all fuels (fossil and biomass) and primary (hydro and nuclear) electricity in 2015. Four and a half billion years ago, as the Earth was formed, the luminosity of the young sun was about thirty percent less than it is today. In that time, the sun has consumed just 0.03 percent of its huge mass but more than half of the hydrogen in its core. The rest of the solar story is of little concern to our civilization which will cease to exist long before the sun transforms itself, first into a red giant (one hundred times its present diameter) whose energy will melt the planet, and then into a much smaller, highly luminous white dwarf: the sun's epochs are measured in billions of years, the history of complex civilizations has, so far, spanned only about 5,000.

We benefit from a perfect delivery: virtually nothing impedes the solar radiation as it streams through the cosmic void, when it reaches the topmost layer of the Earth's atmosphere its flow amounts to about 1367 W/m². This rate is called the *solar constant*, although monitoring from special satellites has revealed tiny, irregular, short-term fluctuations (up to 0.2 percent from the mean) that could not be previously observed because of atmospheric interference. There is also a more regular, tiny (about 0.1 percent), fluctuation connected to the sun's eleven-year activity cycle.

SOLAR RADIATION

Overall, the sun's spectrum corresponds very closely to a perfect black body, radiating at 6,000 K with the maximum emission close to 500 nm, in the lowest wavelengths of green light (491–575 nm). The visible part of the spectrum extends from 400 nm (deep violet) to 700 nm (dark red) (Figure 4): light's diffraction in a rainbow or by a glass prism shows this beautiful color sequence. Human eyes have peak sensitivity for green and yellow (576–585 nm) light, with maximum visibility at 556 nm (near the end of green). Visible light carries about thirty-eight percent of the energy of incoming solar radiation, less than nine percent comes as ultraviolet (UV, less than 400 nm) radiation, which can neither be seen nor felt, and fifty-three percent is in infrared (IR, more than 700 nm) wavelengths, which include detectable heat (see Figure 5). The radiation that we measure at the top of the atmosphere and the radiation we receive on the ground (insolation) differ greatly, both in terms of overall quantity and in spectral composition. The most important quantitative adjustment is obvious: the

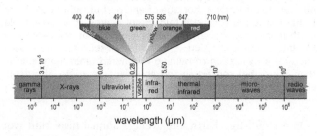

Figure 5 The electromagnetic spectrum

SOLAR RADIATION (*cont.*)

solar constant measures radiation that streams through space and is perpendicular to a flat surface, but because this flow must be distributed over the planet's almost perfect sphere, the average value, per unit area of the rotating Earth, is exactly one-quarter (about 342 W/m²) of the extra-terrestrial flow (the area of a sphere is four times larger than the area of a circle with the same radius). This incoming shortwave radiation is partitioned in three principal ways. Roughly twenty percent is absorbed as it passes through the Earth's atmosphere. The absorption of UV radiation (mostly by stratospheric ozone) accounts for only about a tenth of the overall effect, but the elimination of frequencies below 300 nm was an essential precondition for the evolution of complex life. The remainder is absorbed by tropospheric clouds and aerosols (fine solid or liquid particles suspended in the atmosphere). Global albedo, the share of incoming radiation reflected by clouds and the Earth's surface, without changing its wavelength, is almost exactly thirty percent. Fresh snow and the tops of tall cumulonimbus (thunder) clouds reflect more than ninety percent, dark soils and thick coniferous forests about five percent. About two-thirds of global albedo results from reflections from cloud tops; the remainder is split between reflections from surfaces and back-scattering in the atmosphere.

This means that average insolation amounts to almost exactly half of the solar constant, averaged per unit area of the rotating Earth, or approximately 170 W/m². This insolation adds up to an annual global solar energy input of 2.7 x 10^{24}J, or roughly 87 PW, more than five thousand times the worldwide consumption of fossil fuels and primary electricity in 2015. This makes it obvious that it is not a shortage of energy, but rather our ability to harness it and convert it into useful energy at an acceptable (both monetarily and environmentally) cost, that will determine the fate of our civilization. A tiny share of solar radiation could energize a civilization consuming a hundred times more energy than ours – but converting this abundant flow into affordable electricity is an enormous challenge.

On a cloud-free Earth, the average annual insolation would show a regular poleward decline, but tropical cloudiness causes a notable solar impoverishment of the equatorial zone, and monsoonal cloudiness does the same for the more northerly parts of Asia. Consequently, large parts of equatorial Amazonia, southern Nigeria (just 5° north of the equator), and provinces in the southern half of China (most notably the landlocked Sichuan, situated in a mountain basin at 30°N) receive less sunlight annually than New England, a region that extends from 40 to 45°N. And it is even less appreciated that the peaks of noon summer insolation are virtually identical in Jakarta (Indonesia's capital, located at 6°S), and Edmonton (Alberta's capital, nearly 55°N). These realities have major implications for any future large-scale attempts at direct (photovoltaic) conversion of insolation to electricity.

All the radiation absorbed by the Earth's atmosphere and its solid and liquid surfaces is eventually re-radiated in IR wavelengths, and while the incoming radiation peaks at roughly 500 nm and ninety percent of it is shorter than four μm, the outgoing flux peaks at 9.66μm (a twenty-fold longer wavelength) and extends to just below three μm. This means that the incoming shortwave and the outgoing longwave streams of energy have a small overlap. Three major pathways maintain the Earth's radiation balance: a small amount of energy is returned (through conduction and convection) as sensible heat, and about three times as much radiation leaves as the latent heat of evaporated water, which is released into the atmosphere after the moisture is condensed. Only a small share of the longwave emissions from surfaces (originating from the re-radiation of absorbed shortwave flux and the downward longwave emissions from the atmosphere) goes directly to space: some ninety-five percent is absorbed by the atmosphere's greenhouse gases.

Atmospheric water vapor, the most important greenhouse gas, has several strong absorption bands; at wavelengths between 1 and 8μm, it adds about 20 K to the mean equilibrium surface temperature. Trace (but critically important) concentrations of CO_2 (accounting for about a quarter of the current natural greenhouse effect), methane, nitrous oxide, and ozone increase the surface temperature by more than 10 K. The greenhouse effect has been responsible for maintaining a relatively narrow range of biospheric temperatures for the past 3.5 billion years but water vapor, the main contributor, could not have been the key regulator, because its changing atmospheric concentrations amplify, rather than counteract, temperature changes: water evaporation declines with cooling, and increases with warming. The best explanation involves gradual feedbacks between atmospheric CO_2, temperature, and the weathering of silicate minerals: lower temperatures will bring decreased rates of silicate weathering and result in gradual accumulation of the released CO_2 – and subsequent warming. The key regulatory role played by CO_2 is the main reason for our concerns about relatively large anthropogenic increases of this gas caused mostly by combustion of fossil fuels and also by land use changes (above all by tropical deforestation).

Air and water: media in motion

Absorbed radiation provides three indispensable energy services: it heats continents and oceans (and the heat that these surfaces re-radiate does most of the heating of the atmosphere, keeping it in constant motion), it evaporates water and distributes it far from the sources of its origin, and it energizes photosynthesis. Given the relatively low specific mass of air (one cubic meter has a mass of just 1.2 kg near the Earth's surface, a thousandth of that of water), only a very small fraction of insolation, perhaps no more than two percent, is needed to power the global atmospheric circulation which distributes heat, carries microbes, pollen, and seeds, and is responsible for the wind-driven weathering of continental surfaces. The global atmospheric circulation is energized by the continuous heating of the tropics, which creates a flow of cooler air from higher latitudes toward the equator (creating the so-called intertropical convergence zone) and sets in motion two vigorously moving loops of air commonly known as Hadley cells, after the English physicist George Hadley (1685–1768), who first described their existence.

Warm and humid tropical air first ascends (creating the equatorial low-pressure belt), moves poleward (in both a southerly and northerly direction), then cools, and descends (and is re-warmed) along a broad belt between 25° and 30° of latitude. This subtropical high-pressure belt creates desert zones, and the return flow of warm and dry air toward the equator generates persistent strong trade winds near the ocean's surface. The existence of the trade winds was discovered in 1492, as they carried the three small ships commanded by Christopher Columbus (1451–1506) from the Canaries to the Bahamas in thirty-six days. A weaker circulation is also set off by the outflow of cold polar air that eventually warms up, rises, and returns at higher altitudes to close the loop. The mid-latitude (35°–50°) circulation (the Ferrell cell, named after a 19th-century American meteorologist) is driven by Hadley and polar cells, with air moving poleward near the surface and equatorward at higher altitudes. On a non-rotating Earth, ground winds in the mid-latitudes of the Northern hemisphere would be southerlies but the Earth's rotation deflects them into the prevailing westerlies that bring plenty of precipitation to America and Europe's western coasts.

The fastest near-ground winds are the product of the intensive summer heating that generates cyclonic (low-pressure) flows, ranging from innocuous localized thunderstorms to massive hurricanes. Even big thunderstorms, with a power of ten to a few hundred gigawatts, do not usually produce winds that strike objects with vertical power densities of more than $15\,kW/m^2$, below the threshold for structural damage. North American hurricanes originate off Africa, first move westward, and then veer clockwise, frequently making landfall along the northern Gulf of Mexico, Florida, and the East Coast. Their Asian counterparts (typhoons) originate above very warm Pacific waters near the Marianas, move westward and repeatedly affect large parts of Southeast Asia, coastal China, the Korean peninsula, and Japan. Hurricanes or cyclones can have speeds up to 90 m/s (more than 300 km/h) and they strike vertical surfaces with power densities of up to 1 MW/m², forces easily resisted by modern steel and concrete structures but not by wooden-framed houses.

While some hurricanes can endure for weeks and affect sequentially large areas along a path extending for several thousand kilometers, tornadoes are more restricted. The average path of an American tornado is only about 125 m wide (and often sharply delineated; a barely damaged house can stand across the street from a completely destroyed structure) and less than 10 km long, and they last less than three minutes. In contrast, tornadoes in the most violent (and relatively rare) category can generate winds in excess of 100 m/s and can strike vertical surfaces with more power than a typical hurricane.

WATER'S UNIQUE PROPERTIES

Water's high specific heat capacity, 4.185 J/g°C, is several times that of soil and rock, and that is why the temperature of water rises and falls more slowly than that of solid surfaces and why it retains much more heat per unit of volume, making the ocean the world's most massive temperature regulator. An Earth covered mostly by continents would repeatedly swing between high and low temperatures (similar to the oscillations experienced in large deserts). Moreover (as already noted), water has an extraordinarily high heat of vaporization, nearly 2.5 kJ/g at 20°C, which means that a large amount of latent heat can be moved over very long distances in water vapor and released tens, hundreds, or even thousands of kilometers away from its origin.

WATER'S UNIQUE PROPERTIES (*cont.*)

Evaporation draws water from bare soils and is intensified by transpiration (the movement of water from roots to leaves and then to the atmosphere) from vegetation but, obviously, the ocean dominates the Earth's energy balance, not only because of its extent (just over seventy percent of the planet's surface), but also because its low albedo (on average, six percent) means that it absorbs nearly four times more insolation than the continents. But because of water's poor conductivity (less than one percent of that of even a poorly conducting metal) an inevitable consequence is the ocean's strong thermal stratification. Sunlight penetrates only a thin sliver of the ocean's average depth of 3.8 km, from less than 1 m in highly turbid coastal waters that receive massive inputs of silt from large rivers, to about 200 m in the clearest tropical seas. Wind-generated waves mix the water within a similarly thin layer.

The surface temperature of this shallow mixed layer fluctuates daily and seasonally, and can rise to more than 25°C in the tropics. A more pronounced temporary warming takes place periodically in the Pacific Ocean where, normally, the strong trade winds off South America push the surface waters westward, creating cool surface water temperatures and causing the upwelling of nutrient-rich waters, which supports abundant marine life. But when the trade winds weaken, the surface waters off South America warm up, the upwelling is shut down (as is, largely, the fishing) and the westward expansion of warm surface waters extends along the equator to join warm water off Australasia. This recurrent warming phenomenon is known as El Niño and is associated with heavy rains and flooding in Peru and with drought in Australia and Indonesia. Its opposite is La Niña, which occurs when unusually strong trade winds create a larger than usual pool of cool water off the South American coast.

Below the thermocline (the layer of oceanic water where temperature declines rapidly with depth but nutrient concentrations and salinity increase), the water is always uniformly dark and close to 4°C, the point of its highest density. This is yet another property of this remarkable medium: while the density of other substances increases with decreasing temperature, water is at its densest at 3.98°C. This unusual temperature-density relationship makes it possible for fish to survive in northern waters, as ice forms at the surface rather than the bottom.

WATER'S UNIQUE PROPERTIES (*cont.*)

The cold waters of the deep ocean are brought to the surface only in restricted upwelling zones along the subtropical western coasts of the continents. This upwelling is compensated for by downward convection in giant oceanic cataracts that transfer surface waters to depths of several kilometers.

The planetary water cycle (evaporation–precipitation–runoff) moves, annually, nearly $580,000 \, km^3$. This equates globally to an average precipitation of about 3 mm per day, or 1.1 m a year, for every square meter of the Earth's surface. Some 46 PW are needed to vaporize that mass of water, an energy that amounts to about fifty-two percent of total insolation. Latent heat thus greatly surpasses the kinetic energy of the cyclonic flows that bring summer rains: in thunderstorms the difference is commonly fifty- to one hundred-fold, in hurricanes the heat released during condensation is several thousand times the kinetic energy of the massive moving cyclone. But even the largest hurricane is an insignificant bearer of tropical heat, compared to Asia's summer monsoons, which annually affect almost half of humanity and dump about $10,000 \, km^3$ of rain, from coastal Oman in the west to the Philippines in the east, releasing about five hundred times more latent heat than the most powerful hurricanes.

Only a small part of continental precipitation replenishes deep aquifers: about three-fifths are evaporated and less than a third is returned to the ocean by streams. Given the average continental elevation of 850 m, this stream flow has annually about 400 EJ (13 TW) of potential gravitational energy, an order of magnitude above the world's total electricity use at the beginning of the twenty-first century. Only a small share of this enormous potential can be harnessed by building hydroelectricity generating stations; the degree of exploitation is limited by the availability of suitable sites to locate large dams, by competing needs for water (particularly for irrigation, cities, and industries), and by the necessity to maintain minimum streams flows to support aquatic life and flush away silt.

The earth's heat: refashioning the planet

The other flow that energizes our planet is puny in comparison with solar radiation but its qualitative impact on the evolution of life and its effects on the fortunes of civilizations have been immeasurable, because the Earth's internal heat is constantly recreating the ocean floor and re-assembling and splitting the continents. These grand geotectonic processes are accompanied by repeated, and often catastrophic, disasters. There are two sources of this internal energy: the basal heat from the slow cooling of the Earth's molten metallic (largely iron) core and that from radioactive decay (particularly of uranium 235 and 238, thorium 232, and potassium 40). The latter flux is more important, and while the definite partitioning of the heat's origins is still impossible, we have plenty of measurements to enable us to conclude that the aggregate global power of this geothermal energy amounts to some 44TW.

Dividing this total by the Earth's surface area gives a mean global flow of less than $90\,mW/m^2$ compared to $170\,W/m^2$ for average insolation, a difference of three orders of magnitude. The geothermal flux shows considerable spatial variation: the average for the ocean floor is more than seventy percent higher than that for the continents, where the areas of ancient crustal rocks (the Canadian Shield is a prime example) have the lowest rates. In contrast, the youngest sea floor oozes heat at a rate roughly three times as high as the oceanic average. The highest recorded large-scale averages are along the ocean ridges, where new floor is being created by upwelling of hot rocks (which is why the

Pacific accounts for roughly half the Earth's heat flow). Spectacularly high figures are reached at these hydrothermal vents, which spew water as hot as 360°C and reach power densities of many megawatts per square meter, rates equaled only by major volcanic eruptions.

About sixty percent of the Earth's heat is converted into the formation of new sea floor along some 55,000 km of the ocean ridges, which divide the Earth's crust (its thin, solid, topmost layer) into rigid and slowly moving geotectonic plates (Figure 6). The Pacific plate, the largest, is purely oceanic and in places is less than 10 km thick, while other plates have piggy-backing continents and crustal thicknesses of more than 100 km. Basaltic magma, rising from the underlying mantle along the ridges, creates about three square kilometers of new ocean floor at the mean global spreading rate of less than 5 cm/year. In a few places around the world – the Afar region in Eritrea and Ethiopia, the East African Rift Valley, and central Iceland – the rifting process can be seen on land. Diverging oceanic plates must eventually collide with more massive continental plates and the ocean

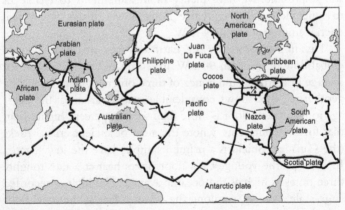

Figure 6 Geotectonic plates

floor must be recycled (subducted) back into the Earth's mantle, a solid, nearly 3,000 km thick, layer between the crust and the liquid core. Deep ocean trenches are the most spectacular features of plate subduction.

This continuous recycling explains why there is no ocean floor older than about 200 million years (most of it is younger than 100 million years), and why the most violent earthquakes (often causing massive tsunami) and volcanic eruptions are concentrated along the subduction fronts. These zones make a huge semi-circle of deep ocean trenches around the Pacific, from the Aleutian Islands to Tonga (north of New Zealand), as the relatively rapidly moving Pacific plate is forced under the virtually immobile Australian and Eurasian plates. The other major type of collision between oceanic and continental plates results in the formation of prominent mountain ridges: the Himalaya is still growing slowly as the Indian plate collides with the Eurasian, and a largely spent collision of the African plate with the westernmost region of the Eurasian plate created the Alps.

Many details about the energetics and mechanics of the planet's grand geotectonic process remain unclear, but there is no doubt that magma upwelling along the ridges and plate subduction along the trenches drive the Earth's most massive cycle. New ocean floor, created by the convection of mantle magma is, on average, about 3 km above the abyssal plain, forming massive blocks of hot rocks with large gravitational potential energy, which furnish the push-power away from the ridges. Along the trenches, the sinking of the cold ocean floor produces the pull-power, as it applies torque to the viscous mantle. The importance of this force is attested by the fact that the average speed of plate movement correlates best with the length of subduction zones: the Pacific plate has short-term generation rates of up to 20 cm/year and long-term velocity of up to 90 km per million years. These speeds prove that mantle drag force (proportional to a plate's area and its velocity) must be relatively small.

EARTHQUAKES AND TSUNAMI

As even the fastest moving plates travel at only about 0.5mm a day, their continuous displacement cannot be perceived directly, but earthquakes and volcanic eruptions remind us repeatedly of the incessant energy flow from the Earth's mantle. All but five percent of earthquakes are associated with subduction or collision zones, and all but ten percent take place in or near the Pacific's coastal areas, in the appropriately named "Ring of Fire". The energy released annually by earthquakes is equal to no more than one to two percent of the total geothermal flux but that is a process of continuous heat convection, while most earthquakes last only between a few seconds and half a minute, meaning that the larger ones have a great deal of destructive power. Consequently, in the twentieth century, earthquakes claimed more lives than volcanic eruptions, cyclones, and floods combined.

The easiest way to find an earthquake's energy is through its relationship with the magnitude of tremors. A standard measure was introduced by Charles Richter (1900–1985) in 1935. Richter's magnitude is the logarithm to the base 10 of the largest trace amplitude (measured in micrometers), recorded with a standard torsion seismometer 100km from the tremor's epicenter (the ground above the focus of the earthquake). The conversion to total energy, released as seismic waves, yields, as do other methods, only approximate values. The largest recorded earthquakes, with a Richter magnitude of 9.0, release nearly 1.5 EJ of energy, and if they take place in less than 30 seconds, the power is as high as 50 PW: no ephemeral energy discharge originating on Earth can be as powerful. At the same time, there is no strong correlation between an earthquake's power and the overall death toll: residential density and, above all, the quality of housing construction are the key determinants of casualties. As a result, two of the twentieth century's most famous earthquakes ended up with very different tolls: the 1906 San Francisco earthquake was roughly four times more powerful than the 1923 Tokyo quake, whose death toll (at nearly 143,000) was (mainly because of the collapse and burning of densely packed wooden houses) almost fifty times higher. The most deadly earthquake in recent history struck on July 28, 1976, in Tangshan, a large coalmining town in China's Hebei province: its magnitude was 7.8 and it killed, officially, 242,219 people in the city and its surroundings but the real toll was much higher.

EARTHQUAKES AND TSUNAMI (cont.)

Some underwater earthquakes generate tsunami, massive seismic sea waves that can travel in the deep ocean at more than 600 km/h, while causing only a minimal disturbance at the surface. Once these waves hit shallow coastal waters they may rise to heights of up to several tens of meters and can strike shoreline vegetation and structures with vertical power densities of more than $1 MW/m^2$, rivalling, and often surpassing, the power of the fiercest tornadoes. The Pacific Ocean has the highest frequency of tsunami and Japan saw most of the tsunami-related casualties in modern history. The tsunami that struck the beaches of Honshū on June 15, 1896 was up to 30 m high and killed 27,000 people. The March 2011 earthquake off the coast of northern Japan had a magnitude of 9.0 and the tremor caused a huge tsunami that was responsible for most of the nearly 16,000 deaths and more than 2,500 missing people. But the most destructive recent tsunami was triggered by an undersea earthquake (magnitude 9.0), centered just off the northwestern tip of Sumatra's Aceh province on December 26, 2004: it killed more than 200,000 people, mainly in Aceh but also along the eastern coasts of Sri Lanka and India, on western beaches of Thailand and (in much smaller numbers) right across the Indian Ocean, in Somalia. The overall energy release associated with this subduction-generated earthquake was estimated at 2 EJ.

The northwestern tip of Sumatra was also the site of one of the largest modern volcanic events: in 1883 a series of eruptions, culminating on August 26, destroyed most of Rakata, a small island in the Sunda Strait surmounted by the cone of Krakatoa, lifting about $20 km^3$ of ash and rocks into the atmosphere. Subsequent tsunami, rather than the eruption itself, caused most of the estimated 36,000 casualties.

Because they are intermittent and often of short duration, volcanic eruptions account for only a small share of the global release of geothermal energy: the best estimates put the share at around two percent of the total flux but there is enormous year-to-year variability, as decades may elapse between spectacular large-scale volcanic events, and some volcanoes erupt violently but briefly while others remain active for extended periods of time.

Heat nearly always dominates the overall release of volcanic energy (by one to three orders of magnitude) but its principal carriers are different. For many volcanoes, the heat is carried mostly by massive ash clouds that rise all the way to the stratosphere (most recently during the eruption of Mount Pinatubo in the Philippines on June 15, 1991), and cause lower ground temperatures worldwide for months. In contrast, Hawaiian volcanoes release their heat in the form of slow-moving lavas (smooth, rope-like, *pahoehoe* and crinkly *aa*); some of these flows can be closely approached to sample the hot magma. By far the most dangerous heat release are the pyroclastic flows that combine volcanic material ranging from fine ash to large rocks with hot (even above 500°C) gases. They can flow downhill at more than 100 km/h and smother everything in their path up to 100 km away. In 1902, such *nuées ardentes* (glowing clouds) killed 28,000 inhabitants of St. Pierre on Martinique after Mount Pelée erupted. In August 1997, pyroclastic flows destroyed a large part of the Caribbean island of Montserrat and they have been repeatedly observed at Unzen in Japan.

The eruption of Mount St. Helens on May 18, 1980 was the best monitored event of its kind. The total energy release over nine hours was about 1.7 EJ (52 TW). The best estimates for other large modern eruptions are: Krakatoa, 1.7 EJ, Bezymyannyi (Kamchatka), in 1956, 3.9 EJ, Sakurajima (Japan), in 1914, 4.6 EJ, and Tambora, in 1815, 8.4 EJ. But even Tambora's energy was pitiful compared to the Yellowstone eruption, 2.2 million years ago, that released an estimated 2,500 km^3 of ash. Even that was a minor event, compared to eruptions spread over some five million years (between 65 and 60 million years ago) that piled up about one million cubic kilometers of basalt lava to form the extensive (about 1.5 million km^2) Deccan Traps in west central India.

Much like earthquakes, volcanoes are overwhelmingly associated with the margins of tectonic plates, but at a few locations powerful hot magma plumes have pierced through a plate, creating spectacular hot spots far away from any subduction or collision zones. The most famous example is the chain of Hawaiian Islands that extends, in the form of seamounts, all the way to Kamchatka. This volcanic chain is being created by a massive hot spot, that keeps piercing the Pacific plate on its northwestern-ward journey and is now situated underneath the western coast of Hawaii (where it manifests itself in continuing eruptions of Kilauea volcano) and just offshore from the island where Loihi, a large undersea volcano, will emerge (but not for tens of thousands of years) as the chain's newest island. Another major hot spot pierces Africa's plate in the center of the continent, creating the Virunga volcano chain on the borders of Uganda, Rwanda, and Zaire (the world's last mountain gorillas live in the bamboo forest in the foothills of one of these volcanoes, the 4,500 m Mount Karisimbi).

Photosynthesis: reactions and rates

Photosynthesis is energized by the absorption of light by pigments in the thylakoid membranes inside bacterial and plant chloroplasts (the cellular organelles that give plants their green color). The energy efficiency of the conversion of simple inorganic inputs into new phytomass is surprisingly low. Introductory textbooks often outline the entire process in a simple equation in which the reaction of six molecules of CO_2 and six molecules of water produces one molecule of glucose and six molecules of oxygen: $6CO_2 + 6H_2O = C_6H_{12}O_6 + 6O_2$. The reality is vastly more complex. The key sequential steps were revealed for the first time in 1948 by Melvin Calvin (1911–1997) and his co-workers (Calvin received the 1961 Nobel Prize for Chemistry for this discovery). Most importantly, the process entails not only carbon fixation and oxygen evolution, but is also the complex exchange of oxygen and CO_2 in two closely related cycles (the other being photorespiration).

Chlorophylls a and b, the two dominant plant pigments that can be excited by radiation, have rather narrow absorption maxima, the first being between 420 and 450 nm, the second between 630 and 690 nm. This means that photosynthesis is energized, overwhelmingly, by a combination of blue and red light, and because the pigments absorb virtually no light in the green and yellow parts of the visible spectrum those colors, reflected from the leaves, dominate in spring and summer and change only as the pigments begin decomposing in the fall. It also means that photosynthetically active radiation (PAR) amounts to only about forty-three percent of insolation. The energy absorbed by the pigments drives the electron transport (water is the electron donor and hence the source of oxygen) that involves three multi-enzyme complexes. This results in the production of NADP (nicotinamide-adenine dinucleotide phosphate, one of the two most important enzymes in cells) and ATP (adenosine triphosphate) which drive the incorporation of CO_2-derived carbon into carbohydrates, a process that follows three distinct paths.

PHOTOSYNTHETIC PATHS

The proper name of this process is the reductive pentose phosphate (RPP) or Calvin-Benson cycle. In the first step, one of the biosphere's most abundant enzymes, ribulose 1,5-bisphosphate oxygenase (commonly known as Rubisco, it accounts for about half of all soluble protein in leaves) catalyzes (increases the rate of) the addition of CO_2 to a five-carbon ribulose 1,5-bisphosphate (RuBP) to form the three-carbon 3-phosphoglycerate (PGA). In the second step, NADPH (NADP with one hydrogen atom added) and ATP produce 1,3-bisphosphate (triose phosphate). Finally, the Rubisco is regenerated and the triose phosphate used either to form carbohydrates or fatty acids and amino acids.

PHOTOSYNTHETIC PATHS (*cont.*)

Rubisco acts not only as a carboxylase (an enzyme which catalyzes the addition of CO_2) but also as an oxygenase (an enzyme which catalyzes the addition of oxygen). In that role, it catalyzes the binding of oxygen to RuBP, leading to the production of PGA (which re-enters the RPP cycle) and glycolate (a two-carbon compound) whose breakdown releases CO_2. Because the atmosphere contains so much more oxygen than carbon dioxide (20.95 percent versus 0.04 percent), the Rubisco-mediated photorespiration cycle (essentially an oxygen-consuming, carbon dioxide releasing counterpart of photosynthesis) can reconvert significant amounts of carbon (as much as half) and thus reduces the net efficiency of photosynthetic conversion. The numbers of carbon atoms in the final products of the two cycles give them their common name, the C_3 (photosynthetic) and C_2 (photorespiration) cycles (Figure 7). Unfortunately, most widely cultivated staples, as well as most vegetable and fruit species, use the C_3/C_2 cycle, giving the so-called C_3 plants (which include rice, wheat, barley, rye, all tubers, all leguminous grains, and all oil crops) inherently low energy conversion efficiencies. Their cultivation also demands a great deal of water.

There is another photosynthetic path, which deviates in an important way from that dominant sequence. Its first step is the hydration of CO_2 to bicarbonate (HCO_3) in mesophyll cells (large cells

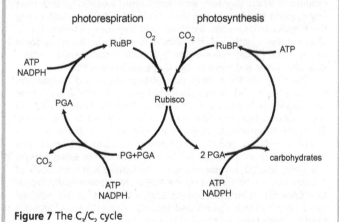

Figure 7 The C_3/C_2 cycle

PHOTOSYNTHETIC PATHS (*cont.*)

that communicate, via stomata (holes), with the leaf's surface); then, rather than using Rubisco, the bicarbonate is reduced, with a different enzyme, phosphoenolpyruvate carboxylase (PEP), to produce oxaloacetate, a four-carbon acid, which is subsequently transformed to malate (also with four carbons). This compound is moved into bundle sheath cells (buried deep inside the leaf, surrounding the vein), where the CO_2 is removed. The released CO_2 enters the RPP cycle, to be fixed by Rubisco. This C_4 cycle is inherently more efficient because PEP carboxylase is a better catalyst than Rubisco and also because oxygen levels in bundle sheath cells (which exist only in C_4 plants) are lower than in mesophyll cells, and hence Rubisco's capacity to act as an oxygenase (as in the photorespiration process) is virtually eliminated while it catalyzes the C_3 cycle.

C_4 species have three more photosynthetic advantages: while C_3 species are saturated by light inputs with power densities around 300 W/m², there is no light saturation in C_4 plants; while C_3 plants perform best at temperatures between 15 and 25°C, C_4 plants have their highest net photosynthetic rates between 30 and 45°C; and while C_3 plants transpire, on average, around 1,000 moles of water (and as much as 4,000 moles) for every mole of CO_2 they import, this ratio is just 400–500 for C_4 plants. These three attributes mean that C_4 plants thrive in sunny climates and are much more heat- and drought-tolerant (because inherently more water-efficient) than any C_3 plant growing in the same conditions. Unfortunately, only three major crops – corn, sorghum, and sugarcane – follow the C_4 photosynthetic path, which is also, unfortunately, shared by some of the world's most invasive weeds (including crabgrass, a particularly obnoxious and ineradicable C_4 species in gardens).

The final photosynthetic path, crassulacean acid metabolism (CAM), is restricted to succulents belonging to the Crassulaceae and Cactaceae families and to some orchids and bromeliads. This process is an adaptation to extreme aridity and high temperatures. These plants absorb CO_2 during the night, when their water losses can be minimized, and convert it into malate using PEP produced from starch. During the day, NADPH and ATP are produced by light reactions, the CO_2 is removed from the malate and high levels of cellular CO_2 are converted by the RPP cycle. Commercially important CAM plants include pineapple, aloe (famous for its medicinal uses), opuntia (cactus pear), and agave (the sweet juice of agave is used to make Mexican tequila) and vanilla.

No matter which path a plant follows, it will consume a significant share of its fixed carbon in respiration, which produces energy by oxidizing sugars. This process takes place in special organelles (mitochondria) within plant cells, and the energy is used to maintain basic functions. These include the transport of sugars or starches (the photosynthate) from leaves to stems and roots, the uptake of macro- and micro-nutrients from soil, and their assimilation into organic compounds. The energy obtained through respiration is also used to make the complex organic compounds needed by organisms for their metabolism, supporting structures (roots, stems, trunks) and defense against heterotrophs, particularly insects. This last is accomplished by making the soft interior tissues inaccessible behind thick barks, waxy leaves, or thorns. Respiration may claim less than a fifth of all new photosynthate in crops (humans provide readily available nutrients through fertilization and defense with insecticides) but uses all of it in mature trees, which spend their energy completely on the maintenance of existing structures, as their newly synthesized phytomass is used to replace ageing parts, rather than form new tissues.

The maximum theoretical net efficiency of photosynthesis (after subtracting all respiration losses) is about four percent of insolation, but this rate can be approached only during brief periods and in the presence of adequate water and nutrients. Intensively tended (irrigated, fertilized) crops can average two percent efficiency during their growing season; the most productive temperate and tropical forests approach 1.5 percent. The global continental average is only 0.33 percent, and, because oceanic phytoplankton converts less than 0.1 percent of insolation into new aquatic phytomass, the average for the entire biosphere is less than 0.2 percent. To put it differently, the energy of only one in 500 photons which reach the planet's ice-free surface gets converted into new phytomass. Given the immense flow of solar radiation, this inefficiency is largely irrelevant; the overall photosynthetic performance is impressive in quantitative terms and even more so in qualitative differentiation. The best global calculations show an annual net primary production (NPP) of about 120 billion tonnes of plant mass on the continents and 110 billion tonnes in the ocean.

Forests are generally the most productive ecosystems. The distribution of major biomes (complex large-scale communities of organisms) is limited not by insolation but primarily by temperature and precipitation: tropical rainforests need at least one meter of rain a year and an average annual temperature of more than 20°C; deciduous forests, the dominant natural vegetation of Western Europe and Eastern North America, occupy a broad niche, with annual average temperatures between 0 and 20°C and rainfall between less than half a meter and more than two meters a year. The annual rates of NPP (in absolutely dry matter) range widely for major ecosystems: from $1–3.5 \text{kg/m}^2$ (10–35t/ha, i.e. tonnes/hectare) in tropical rainforests, to $0.5–2.5 \text{kg/m}^2$ in temperate forests and $0.2–1.5 \text{kg/m}^2$ for most grasslands.

In qualitative terms, the richest tropical rainforests, in Amazonia, harbor more than six hundred different species per hectare, but most of this phytomass is stored in a relatively small number of large canopy and emergent (towering above the general canopy level) trees. Temperate and boreal forests are often dominated by just a few species, but some of these ecosystems, particularly the rainforests of the Pacific Northwest, can store much more wood per hectare (up to 3,500t) than their richest tropical counterparts. But the average phytomass in the above-ground vegetation of temperate forests (root phytomass is always difficult to assess, and so most phytomass totals are limited to above-ground growth) is similar to that stored in tropical formations, from 250 to 300t/ha. There is no universal conversion rate of these masses to energy: woods with higher lignin and extractives (resins, waxes) content will have higher heating values. The range for common

North American species is from 17.8 MJ/kg for sweetgum to 21 MJ/kg for Douglas fir.

NPP rates are essentially equal to annual yields for crops that have not suffered any major damage from pests and diseases. These rates are, predictably, highest for C_4 crops and for crops grown under optimum conditions. Many Iowa farmers harvest more than 12 t/ha of corn, excellent harvests of English or Dutch wheat are around eight, and in Japan and China's coastal provinces rice produces just over six. Leguminous crops (beans, lentils, peas) yield mostly less than two, but many vegetables can produce in excess of fifty. Because yields are reported at harvest moisture (which is less than fifteen percent for cereal and leguminous grains but ninety to ninety-five percent for many vegetables) the latter reduce to less than 5 t/ha of absolutely dry matter. Similarly, harvested stalks of sugarcane (excellent yields of this C_4 plant are from 80 to 100 t/ha) contain only about twenty-five percent dry matter. Well-managed forests have (because of considerably higher respiration rates) annual increments of between 1 and 2 t/ha of dry matter but intensively managed plantations of fast-growing tree species (poplars, eucalyptus, pines) can produce more than three or four t/ha.

Heterotrophs: metabolism and locomotion

Heterotrophs have two basic ways of metabolizing complex organic compounds: anaerobically (without the presence of oxygen) and aerobically. The two groups of anaerobic fermenters that figure prominently in human affairs are the bacteria responsible for lactic fermentation and yeasts. Fermenting bacteria, which transform sugars into lactic acid, give us sour milk, yoghurt, and cheese, as well as sauerkraut, gherkins, and olives. Yeasts (fungi belonging to the phylum Ascomycota) produce alcohol from a variety of sweet or starchy substrates. The anaerobic path arose during the early history of the Earth, when its atmosphere had only a trace of oxygen; the aerobic became widespread after the oxygenating atmosphere was created during the past two billion years.

In the presence of oxygen, yeasts do not produce alcohol but carbon dioxide, to make dough that rises: leavened bread instead of pita or chapati. About half of the bacterial phyla are aerobic; they include commercially important species, such as the nitrogen-fixing bacteria (above all those of the genus *Rhizobium*) that provide leguminous species with this key nutrient (in the form of ammonia) in return for their supply of carbohydrate. All the species belonging to the kingdom Animalia are aerobic heterotrophs. The first task of heterotrophic metabolism is to break down carbohydrates into their constituent monosaccharides (glucose and fructose; sucrose, the world's leading sweetener, is a disaccharide made by their combination), to hydrolyze (react in the presence of water) lipids into glycerol and fatty acids, and to decompose proteins into the constituent amino acids used to build new protein structures (muscles, organs, cells). ATP conserves the energy released by the degradation of nutrients; its transformation to adenosine diphosphate (ADP) makes energy available for the synthesis of new biomass and for locomotion.

Numerous enzymes catalyze these complex reactions. In fermentation, the final products are either lactic acid or ethanol and CO_2 (we keep the gas in champagne but not in table wines) while aerobic metabolism produces, besides large amounts of ATP, CO_2 (which we exhale) and water (which we lose through breathing and in sweat, urine, and feces). The maximum over-all efficiency of anaerobic fermentation is around thirty percent, of aerobic metabolism about sixty percent, both for converting simple sugars or fatty acids. Every heterotroph has its own basal metabolic rate (BMR), the constant minimum power needed to energize its vital internal organs.

BASAL METABOLIC RATE

This rate is measured at complete rest, several hours after the last intake of any nutrients (digestion increases metabolism) and in a temperature-regulated setting. Oxygen consumption (or CO_2 generation) served for decades as its best determinant. In 1932, Max Kleiber (1893–1976) noted that the basal metabolic rates of mammals, ranging from rats to steer, depended only on their body weight raised to the power 0.74. Later, he opted for the exponent of 0.75 (hence the so-called ¾ power law). When body mass (w) is expressed in kg and power in W then BMR = $3.52w^{0.75}$. When both variables are plotted on a logarithmic scale the rates are on, or very close to, a straight line (Figure 8). BMR measurements eventually extended the mammalian line from shrews to elephants and confirmed that its 3:4 slope also

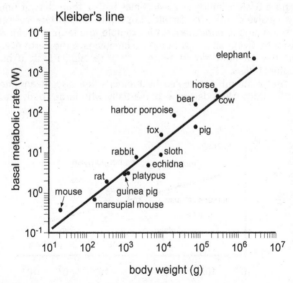

Figure 8 Kleiber's line (plotted from data in Kleiber, M. 1961. *The Fire of Life*)

BASAL METABOLIC RATE (*cont.*)

approximates the rates for some birds, some ectotherms (higher organisms that do not actively regulate their core body temperature) and many micro-organisms.

At close to 0.9, the exponent is considerably higher for carnivorous mammals, which means that their BMR is greater than that of similarly massive non-carnivores, and that it increases more quickly with greater body mass, making a rhino-sized tiger (1 t rather than 100 kg) impossible. On the other hand, the exponent is much lower (just below 0.5) for desert rodents, an adaptation that minimizes their energy needs (as well as water consumption) in a hostile environment. And while the BMR of many endotherms (vertebrates that actively maintain their core body temperature very close to a constant) and ectotherms may share the same exponent, endotherms must pay for their thermoregulation (37–38°C in mammals, and 38–42°C in birds) by having a BMR as much as 20–40 times higher than that of similarly massive ectotherms. Smaller, but significant, differences can be found among endotherms: for example, the large and almost constantly flying albatross (some circumnavigate the Earth every year) have a BMR almost twice as high as similarly-sized but occasional fliers.

The relationship revealed by Kleiber's line also means that specific BMRs (W/g) decline exponentially with larger body mass

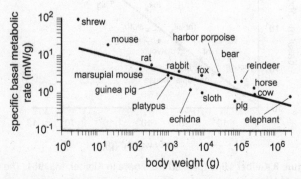

Figure 9 Specific BMR

BASAL METABOLIC RATE (*cont.*)

(at a rate of roughly $w^{-0.25}$): the smallest shrew needs about 100 mW/g, a nearly five orders of magnitude heavier steer requires only 1 mW/g, and a rufous hummingbird, the smallest (3.5 g) avian flier in Canada, has a BMR about twenty times higher than a hundred times heavier (3.5 kg) brown pelican (Figure 9). This limits the mass of the smallest endothermic organisms, because otherwise animals lighter than shrews and hummingbirds would have to feed constantly in order to compensate for high rates of heat loss, a consequence of having a relatively large body surface in relation to the body mass. At the same time, the low specific BMR of large mammals makes it possible for them to go for days without eating and, when some of them lower their BMR and stay at rest, to hibernate (live off the accumulated reserves of fat) for months (as do bears).

The overall energy expenditure of heterotrophs with limited mobility (many invertebrates, and some reptiles and fish) is only a modest multiple of their BMR, while vigorous fliers and widely roaming mammals need most of their energy for their demanding activities. Observations over longer periods of time indicate that the total daily metabolism is about 2.5 times BMR for rodents and more than four times BMR for birds who forage on the wing. Running, flying, and swimming speeds, and endurance are limited by the metabolic scope (the ratio between short-term maximum energy expenditure and BMR) of heterotrophs. The average scope for mammals (independent of body mass) is about ten, for some fish and most birds about fifteen (but avian BMR is much higher than that of fish or mammals!), and commonly less than five for reptiles and amphibians. As a result, songbirds can develop up to 150 mW/g from their tiny bodies, while small rodents put out around fifty and an iguana less than five.

But running mammals give many false impressions: hares seem to be always running quickly, while canids (wolves, wild dogs, coyotes, jackals) often just amble. Whilst hares have a metabolic scope of around eight, canids have the highest known, surpassing thirty. They are the best sustained runners of the animal world: cheetahs may be record holders (105 km/h, greyhounds can do only fifty-eight) but their rushing distance while hunting is usually less than one hundred meters, and they often stop, panting with exhaustion, after running just for ten seconds. The slower wolves can pursue their prey non-stop for twenty minutes and cover, daily, more than thirty kilometers. And size matters, because the power available for running increases more quickly with larger body mass than does the cost of running. As a result, a massive animal may look too bulky and too muscle-bound to be a sprinter, but this could be a deadly miscalculation: you cannot out-run a bison, grizzly bear, or hippo!

If you are a good sprinter you may outrun an Asian elephant (although your chances of encountering one in the wild are now, unfortunately, negligible). These massive animals can do nearly 7 m/s (about 25 km/h), although it is still not fully clear if their rapid locomotion amounts to true running, so if you can run a hundred meters in less than fourteen seconds (the world record is 9.58s) you can elude them. Measurements show no major difference between the cost of running bipedally and quadrupedally, but besides their greater power, larger animal runners can also take advantage of the elastic strain that is stored, temporarily, in their long leg muscles and tendons and then released as elastic recoil: at high speeds this ability, which humans share with horses and cheetahs, can greatly reduce the overall energy needs of running.

Not surprisingly, comparisons of energy cost for the three main forms of locomotion show that swimming, thanks to water's buoyancy, is the least costly way to move: it requires less than one-tenth the energy of running; while running is more energy-demanding than flying for body weights up to 1 kg. For heavier birds, the difference narrows, and the mass of the largest fliers is limited by the fact that the power needed for flight rises more quickly than the power that can be delivered by the pectoral muscles. Consequently, large birds have difficulty getting airborne (they often launch themselves into a strong wind from cliffs or steep hillsides: for example, American condors and royal albatrosses). There are very few fliers heavier than ten kilograms, the heaviest, the East African Kori bustard (13–19 kg), gets airborne only rarely, in a series of low-level hops.

The long-distance migrations accomplished by running, swimming, and flying are much easier when the animals can feed along the way, as do caribou, salmon, and Canada geese. But, counter-intuitively, songbirds migrating in cool weather may spend more energy in stopovers than during their interrupted flight. By far the most spectacular long-distance migration is the non-stop flight of tiny songbirds, as they cross forbidding expanses of ocean on their way from northern and temperate nesting habitats to subtropical and tropical winter sites. These migrations (guided by stars, sun, geomagnetic field, and polarized light) must be powered by fat, the most energy-dense fuel a body can store. Consequently, the maximum distance (discounting any helping winds) a small songbird will be able to fly non-stop will be limited by the amount of fat it can store before it embarks on a trip: many species add as much thirty percent to their normal body mass and have virtually no reserves left when they land (or literally drop exhausted on beaches) at their distant destination.

Energy in ecosystems: networks and flows

Thermodynamic imperatives rule the energy flows through ecosystems. Photosynthesis may be a relatively inefficient process, but autotrophs are at the top of the biosphere's cascade of energy conversions, and hence their phytomass must always dominate the total biomass of an ecosystem. The more steps removed a living thing is from the primary energy of the Sun, the less energy is available at the successive feeding (trophic) levels (positions along what is often, though erroneously, referred to as a food chain – it's almost always a web).

No scientific studies are needed to know that herbivores (primary consumers such as rodents and ungulates) are considerably more abundant than secondary consumers (carnivores, be they birds of prey that eat rodents, or ladybugs that eat aphids). These, in turn, are more numerous than tertiary consumers (secondary carnivores such as aquatic birds that feed on frogs that eat insects or on carnivorous fish). And, needless to say, access to food is greatly enhanced by being an omnivore and eating, indiscriminately, any available autotrophs, heterotrophs, or decomposing biomass – as long as it can be easily found, caught, and digested.

Omnivory is very common, and often is a matter of opportunity or need, as many normally herbivorous or carnivorous species can move up or down one trophic level, as some resources become temporarily abundant and easy to harvest or others become scarce. Humans, always no-holds-barred omnivores, have mastered this approach, as the foodstuffs we have consumed in times of natural plenty or desperate need range from whale blubber to willow bark.

Finally, decomposers play a critical role in every ecosystem, as they break down complex organic macromolecules and make essential macro- and micro-nutrients available, again and again, both to autotrophs and heterotrophs: they can, of course, feed on dead organisms at any trophic level.

In most terrestrial ecosystems, the feeding cascades are short. In the East African grasslands, paragons of large animal abundance, most transfer ends after just two steps with secondary consumers: ungulates (antelopes, gazelles, impalas, wildebeest) eat grass, felids (leopards, cheetahs, lions) and canids (wild dogs, hyenas) eat ungulates, and decomposers feed on their remains and anything else they can break down with their enzymes. No fierce carnivores can be found feeding at the sixth level from the sun. In tropical rainforests – with a greater, much more varied, standing phytomass and a greater variety of heterotrophs – three levels are common, and five are possible: fungi feed on plants, arboreal invertebrates feed on fungi, frogs feed on invertebrates, snakes feed on frogs, and birds feed on snakes.

Marine ecosystems are based on primary production by phytoplankton, a category of organisms that embraces an enormous diversity of tiny autotrophs, including bacteria, cyanobacteria, archaea (unicellular organisms that are outwardly indistinguishable from bacteria but have a distinct genetic make-up), and algae. Marine food webs are generally more complex than those of terrestrial biomes. They can extend to five and in kelp forests six trophic levels and, in an unmatched complexity the richest coral reefs may go seven.

A complete account of biomass within a unit area of any terrestrial ecosystem would show a pyramid-shaped distribution, with autotrophs forming a broad base, herbivores a smaller second tier, omnivores and first-order carnivores at the next level, and the rare top predators at the apex. The mass of the levels varies greatly among ecosystems, but phytomass is commonly twenty times larger than the zoomass of primary consumers and the zoomass of top carnivores may add up to less than 0.001 percent of phytomass.

In marine ecosystems the pyramid is inverted: the brief lives of phytoplankton (mostly between 12 and 48 hours) and the high consumption rates by zooplankton and larger herbivores mean that the total standing heterotrophic biomass could be between two and four times as large as the mass of the photosynthesizing phytoplankton. What is true collectively is also true individually, as most oceanic autotrophs are species of microscopic phytoplankton (their diameters average only about 10μm) while the organisms typical of higher trophic levels – zooplankton as primary consumers, small fish as secondary, larger fish and common squid as tertiary, and tuna as quaternary feeders – are progressively larger: before overfishing greatly reduced their mass and number mature Southern bluefin tuna could weigh more than 150 kg. There are notable exceptions: the largest marine mammals (blue whales, weighing up to 130 t) and fish (whale sharks, weighing up to 1.5 t) are filter feeders, consuming large quantities of tiny phyto- and zooplankton.

The declining numbers of heterotrophs in higher trophic levels are often associated with increasing body size: top predators commonly include the largest animals in their respective classes, be they golden eagles among the birds of prey, or tigers and lions among the felids. Herbivory has obvious energetic advantages, and in all modern ecosystems the animals with the largest body mass are megaherbivores (grazers with body mass greater than one tonne) such as elephants, hippos, and giraffes in the tropics, and moose and muskoxen in boreal and Arctic environments. This primacy was even more pronounced in the past, when the largest megaherbivores (be they the relatively recently extinct mammoth, or enormous dinosaurians, the heaviest one weighing perhaps more than 80 tonnes) were up to an order of magnitude more massive than today's heaviest species.

Generalizations regarding the transfers within the trophic pyramid have been elusive. Pioneering studies done by Raymond Lindeman (1915–1942) on aquatic life in Wisconsin's Lake Mendota found an efficiency of 0.4 percent for autotrophs, while the primary consumers retained less than nine percent, the secondary about five percent, and the tertiary feeders some thirteen percent of all available energy. These approximations were (erroneously) generalized into the "ten percent law of energy transfer," with a corollary of progressively higher efficiencies at higher trophic levels. Subsequent studies proved that neither conclusion was correct, and showed that bacteria and herbivores can be much more efficient converters than carnivores. There are only two safe generalizations: first, no energy loss in any ecosystems is ever as high as that associated with photosynthesis, and second, energy losses during the subsequent transfers to higher trophic levels are never that large, but net transfers are commonly much lower than ten percent.

ENERGY EFFICIENCY IN ECOSYSTEMS

Final energy transfers in ecosystems are the products of exploitation, assimilation, and production efficiencies. The share of phytomass eaten by herbivores normally ranges from just one or two percent in temperate forests to as much as fifty to sixty percent in some tropical grasslands. This rate ignores occasional spikes caused by infestation of insects: gypsy moths can defoliate large areas of boreal trees and migratory locusts can strip more than ninety percent of available phytomass as they move through North African landscapes. Excluding soil fauna, the transfers are rarely above ten percent in any temperate ecosystem, and are mostly around one percent for vertebrates. It should be noted that the abundance of herbivores is not usually limited by the availability of phytomass, but rather by predation by carnivores, while the numbers of carnivores are generally limited by the abundance of prey they can capture.

Assimilation efficiencies (the share of ingested energy that is actually metabolized) clearly depend on feed quality: they are low (commonly less than thirty percent) among herbivores feeding on often digestion-resistant plant polymers, very high (in excess of ninety percent) for carnivores that consume high-lipid, high-protein zoomass. For many species the final rate, that of converting the digested energy into new biomass, negates the advantages of carnivory. This production efficiency is, regardless of their trophic level, much higher among ectotherms. Invertebrates can convert more than twenty percent, and some insects fifty percent, of assimilated energy into new biomass, while the mean for endotherms is around ten percent, for large mammals no more than three, and for small mammals and birds less than two.

Consequently, the share of energy available at one level that is actually converted to new biomass at the next level above –

ENERGY EFFICIENCY IN ECOSYSTEMS (*cont.*)

variously called trophic, ecological, or Lindeman's efficiency – ranges from a small fraction of one percent for passerine birds to around thirty percent for insects. Moreover, the rates show few clear correlations based on taxonomic, ecosystemic, or spatial commonalities. In any case, trophic efficiency is not a predictor of evolutionary success, as both low-efficiency endotherms and high-efficiency ectotherms have done comparably well in similar ecosystems: for example, in African savannas elephants will harvest annually as much phytomass per unit area as termites.

In complex food webs, it is enough to reduce single energy flow by diminishing the abundance of a single species (be it through a climatic change or disease, or because of a human action) to get some unexpected results. A perfect example, which unfolded during the last quarter of the twentieth century, was the massive damage done to the kelp forests of the Pacific Northwest, and hence to the numerous species that depend on these giant marine plants, by sea urchins. The urchin stock was previously controlled by sea otters, but their numbers declined because of predation by orcas (killer whales). These large, sleek mammals have always preferred bigger prey (such as sea lions and seals) but once they became less available, mainly because of the combined effect of overfishing and climatic change, the orcas turned to otters.

3

Energy in human history: muscles, tools, and machines

Our species has spent more than ninety percent of its evolution in small groups of foragers (gatherers, hunters, and fishers). For tens of thousands of years after leaving Africa, our ancestors lived mostly without permanent abodes, relying on their somatic energy (muscles) and, increasingly, on their reasoning to get their food, defend themselves against wild animals and hostile groups of other foragers, construct better shelters, and produce a variety of simple artefacts.

Human inventiveness and adaptability first manifested itself in the use of fire for warmth, preparation of food, and protection against animals. The earliest stone artefacts were followed by clubs and wooden digging sticks, bows and arrows, and spears and tools carved from bone. These tools magnified the limited capacities of human muscle. The obvious limitations of the preserved record and the uncertainties of dating mean milestones are approximate, but the first use of fire may have been more than 1.5 million years ago (by *Homo erectus*), but first bows and arrows are no older than about 25,000 years, and the oldest fishing nets about half that.

The first fundamental extension of humans' inherently limited somatic capacities came from the domestication of large animals (starting with cattle in around 8500 B.C.E.; horses followed some 3,000 years later). These animals were first used for draft (to pull carts, wagons, and agricultural implements, most notably simple wooden plows) after the development of, at first inefficient, harnesses. Even in those societies where more powerful (better-fed) horses gradually supplanted weaker and slower oxen, most farming tasks still required heavy human labor, with long hours of strenuous physical exertion. This situation changed radically only on the arrival of the internal combustion engine first installed in tractors at the beginning of the twentieth century.

In contrast, many stationary tasks, that were for centuries done by people or animals (from the milling of grain to the pumping of water) began to be mechanized in antiquity. Water-wheels, the first simple mechanical prime movers, converted the power of flowing water into rotary motion. Windmills came later, and both slowly evolved into much more powerful and more efficient machines, used in many mining, metallurgical, and manufacturing tasks. Sailing ships, the earliest with only simple sails and very poor maneuverability, were the only other important converters of indirect solar energy flow into useful motion. As far as the provision of heat is concerned, there was no fundamental change from prehistory to the early modern era (usually dated as 1500–1800): the burning of any available phytomass (in inefficient open fires, fireplaces, and simple furnaces) in its natural state and later in upgraded form (as wood was made into charcoal) supplied all household and manufacturing thermal energy needs.

Even after waterwheels and windmills became relatively abundant in some parts of the Old World, and even after more efficient and larger designs made their appearance during the early modern era, animate energy remained dominant, until the machines of the industrial age diffused in sufficient quantities to become first the leading, and soon afterwards the only, important prime movers. This epochal shift (commonly but wrongly called the Industrial Revolution) began in Western Europe during the eighteenth century but was only totally accomplished throughout the entire continent and in North America by the middle of the twentieth. The transition from animate to inanimate prime movers (and from biomass to fossil fuels) is yet to be completed in large parts of Asia and most of sub-Saharan Africa, where human and animal muscle (and wood and charcoal) remain indispensable.

This chapter opens with a survey of the basic realities of human energy needs and human capacity for work, followed by brief summaries of the pre-industrial progress of energy conversion. The sequence begins with the energetic imperatives that governed foraging (gathering and hunting) societies, then moves to more detailed descriptions of energy uses and harvests in traditional farming and energy sources and conversions in pre-industrial cities and manufactures before concluding with appraisals of large waterwheels and windmills. These machines were not only the most powerful prime movers of the pre-industrial era but were also indispensable in the early stages of industrialization when their power, rather than that of steam engines, energized many mechanized tasks.

Human energetics: food, metabolism, activity

Humans must ingest three kinds of macronutrients – carbohydrates (sugars and starches), lipids (fats), and proteins – and more than thirty micronutrients, which fall into two classes: minerals (such as calcium, potassium, iron, and copper, needed in relatively large amounts, and selenium and zinc, sufficient in trace quantities), and vitamins, (water soluble B complex and C, and compounds that dissolve only in fats: A, D, E, and K). Carbohydrates have always provided most dietary energy in all but a few pre-industrial societies (maritime hunters and some pastoralists were the only notable exceptions), but most cannot be used by humans, as we are unable to digest lignin, cellulose, and hemi-cellulose, the compounds that make up wood, straw, and other cellulosic phytomass.

CARBOHYDRATES, LIPIDS, PROTEINS

Digestible carbohydrates come from three principal sources: cereal grains (rice, wheat, barley, rye, corn, sorghum, and minor varieties including quinoa and buckwheat); leguminous grains (beans, peas, lentils, soybeans, and chickpeas); tubers (white and sweet potatoes, yams, and cassava) and fruits (with scores of tropical and temperate varieties). The digestible energy in these common dietary carbohydrates largely comes from complex starches (or polysaccharides) made up of thousands of glucose molecules, but scores of tropical and temperate fruits supply simpler sugars, the monosaccharides fructose and glucose. Refined granulated sugar, which only became widely and inexpensively obtainable in the nineteenth century, is a disaccharide (sucrose, made of glucose and fructose). All these compounds, be they complex or simple, contain 17 kJ/g. They are consumed in a vast array of (baked, boiled, steamed, and fried) foodstuffs: the world's four leading processed carbohydrate products (by mass) are milled rice, wheat flour, corn meal, and refined sugar.

Lipids (fats) are, with 39 kJ/g, by far the most energy-dense nutrients. Their essential fatty acids are irreplaceable as precursors for the synthesis of prostaglandins (lipids that regulate gastric function, smooth-muscle activity, and the release of hormones), and as carriers of fat-soluble vitamins. The major division of lipids is between plant oils and animal fats. Rapeseed, olives, soybeans, corn, peanuts, oil palm, and coconuts are major sources of plant oils for cooking; butter, lard, and tallow are the three main separable animal fats; lipids that are part of animal muscles (or form their surroundings), or are dispersed in milk, are also digested in the consumption of meat, fish, and dairy products. Through history, typical lipid consumption has gone from one extreme to another: it was very limited in most pre-industrial societies (to the point of deprivation), but has become excessive in many affluent countries.

Proteins (made of amino acids), are used as a source of energy (they contain 23 kJ/g) only if the supply of the other two macronutrients is inadequate: their primary role is as indispensable structural components of new body tissues. Human growth requires

CARBOHYDRATES, LIPIDS, PROTEINS (*cont.*)

a balanced supply of essential amino acids (they cannot be synthesized in the human body), to provide the proteins needed to produce enzymes, hormones, antibodies, cells, organs, and muscles, and to replace some of these compounds and structures as the organism ages. All animal foods (and mushrooms) supply all the essential amino acids in the proportions needed for human growth, while plant proteins (whether in low-protein sources, such as tubers, or high-protein foods, such as legumes and nuts) are deficient in at least one amino acid: for example, cereals are deficient in lysine, and legumes in methionine. Strict vegetarians must properly combine these foodstuffs to avoid stunted growth.

Energy content of nutrients and foodstuffs

Nutrients	MJ/kg
Pure lipids	39.0
Pure protein	23.0
Pure carbohydrates	17.0

Foodstuffs	MJ/kg
Butter	30.0
Ethanol	29.3
Cereal grains	14.5–15.5
Lean meats	5.0–10.0
Fish	3.0–9.0
Potatoes	3.0–5.0
Fruits	1.5–4.0
Vegetables	0.6–1.8

Healthy individuals, consuming balanced diets, can convert macronutrients with high efficiency: ninety-nine percent for carbohydrates, ninety-five percent for lipids, and ninety-two percent for proteins (but only about eighty percent of digested protein is available for tissue growth or activity, as more than twenty percent is lost through urine). The actual energy available for human metabolism, growth, and activity is basically equal to the gross energy content for carbohydrates ($17 \, kJ/g$), only marginally lower for lipids ($38 \, kJ/g$ rather than $39 \, kJ/g$) but appreciably lower for proteins ($17 \, kJ/g$). Food composition tables (still in kilocalories rather than joules) are almost always constructed with these reduced values. Some people consume a great deal of energy as alcoholic beverages (beer and wine lead in total worldwide volume); pure ethanol has a relatively high energy density, at $29.3 \, kJ/g$. Metabolized food energy is converted into new cells and organs with efficiencies as high as fifty percent for infants and about thirty percent for adults.

Naturally, the share of total energy needed for growth becomes marginal after puberty, when the final use is dominated by basal metabolism and the needs of various physical activities; as already noted, mental exertions add only very small amounts to BMR, as liver, brain, and heart account for most metabolic energy, even during sleep (Figure 10). The relationship between body weight and BMR has been determined by extensive measurements of oxygen consumption; linear equations derived from these data sets are reasonably good predictors of individual rates for children and adolescents but give poor results for adults. The BMR of two physically identical adults (same sex, same weight, and same body mass index) commonly differ by ten to twenty percent and the disparity can be greater than thirty percent. In addition, BMR varies not only among individuals of the same population but also between different populations, and the specific rate declines with age (Figure 11).

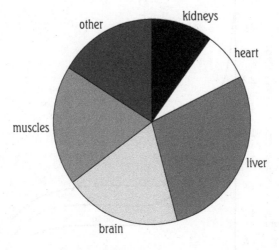

Figure 10 Relative share of BMR in adults (plotted from data in FAO. 1985. *Energy and Protein Requirements*)

As a result, standard BMR-body mass equations, derived overwhelmingly from measurements in Europe and North America, produce exaggerated estimates of energy needs among populations of non-Western, and particularly tropical, adults older than thirty years. Differences in body composition (shares of metabolizing tissues) and in metabolic efficiency are the most likely explanation of this disparity. Keeping this variability in mind, the BMR of adults with body weights between 50 and 80kg (and with desirable body mass index) fall mostly between 55 and 80 W for females and 60 and 90 W for males. Because of their higher share of subcutaneous fat (nonmetabolizing tissue) females always have a lower BMR than males with the same body mass.

The BMR accounts for most of the daily food consumption in all but very active adults. Mark-ups for physical activity

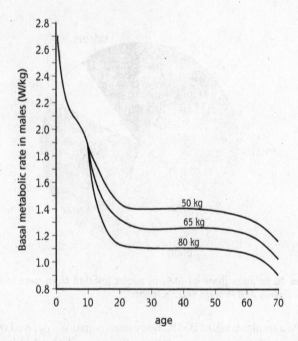

Figure 11 The lifetime progression of specific BMR in men and boys (plotted from data in FAO. 1985. *Energy and Protein Requirements*)

(as multiples of BMR) range from less than 1.2 for housebound people (whose only exertion is to maintain personal hygiene), to less than 1.6 for light labor, less than 1.8 for moderate work, and more than 2 for heavy exertion. In modern mechanized societies, the last category is limited to only a few taxing jobs and leisure activities (for example, lumberjacks and marathon running) but such exertions are still common in traditional farming (the digging and cleaning of irrigation canals is a common example) and they were the norm, although for only limited periods of time, among foragers and in pre-industrial agriculture. These multiples translate into daily energy needs of more than 4,000 kcal for hard-working adult males to less than 2,000 kcal for petite females engaged in light activity. Adult daily needs are most commonly between 2,000–3,000 kcal.

The short-term limits of human performance are far higher: in trained individuals anaerobic outputs can surpass 4 kW, or even 5 kW, for a few seconds; longer (largely anaerobic) exertions (10–15 s) can peak above 3 kW for sprinters and swimmers. Healthy adults can work (aerobically) at a rate that corresponds to a sustained metabolic scope (multiple of their BMR) well above ten. Endurance running is the most natural sustained activity requiring such a high metabolic scope: for untrained adults their maximum ranges from ten to twenty, but for élite athletes (and undoubtedly for many traditional hunters) it would be around twenty-five. This performance translates to as much as 1.75 kW, an equivalent of more than two horsepower (only canids can do relatively better). Despite this relatively high energy cost, humans have several remarkable advantages as endurance runners. Most importantly, our bipedality makes us the only runners that can uncouple the frequency of breathing from our stride. In running quadrupeds, the thorax bones and muscles must absorb the impact of the front limbs, and so the inflation of their lungs is automatically limited to one breath per stride, while our breathing can vary relative to stride frequency. Quadrupedal runners thus have optimum running speeds (it costs them more energy to run both more slowly or more quickly), determined by the structure of their bodies, while humans can run at a range of speeds, and the energy cost of running remains essentially the same for speeds between 2.5 and 6 m/s (9–21 km/h). Our superior thermoregulation is perhaps an even greater advantage for endurance running, as we can get rid of metabolic heat not only by radiation and convection but also by sweating.

HUMAN THERMOREGULATION

As our bodies get warmer, we dilate the peripheral blood vessels and circulate more blood through the superficial veins and then, when skin temperature approaches 35°C, we begin to sweat. When running (or engaged in another form of heavy exertion), we perspire at rates unmatched by any other sweating mammal (most of them, including canids, the superior runners, actually do not sweat, as their panting and tongue-lolling indicate). Horses are moderately good at it, able to lose about 100 g of water per square meter of their surface area every hour, and camels can lose up to 250 g/m², but our common maximum is twice that high! Without sweating, the combination of respiration (water loss from lungs when breathing) and skin diffusion could dispose of roughly 20W of heat in an adult but maximum sweating rates boost this to around 600W. Trained and highly acclimatized individuals (northerners can approach tropical rates of sweating in less than two weeks) can lose heat at a rate surpassing 1,300W.

So as long as we are adequately rehydrated, we can cool ourselves better than any other large mammal, a capacity that enabled hunters to run down desert ungulates (see the next section), miners to work in South Africa's deep gold mines (at more than 3 km below ground their temperatures are above 50°C), and supermarathon (100 km) runners to compete in August heat. But the recent fashion of continuous rehydration (drinking from a plastic bottle every few minutes) is laughably unnecessary: it is a part of our evolutionary heritage that healthy bodies can tolerate periods of moderate dehydration with no ill-effects, as long as the liquid is replenished: marathon runners do not replenish their water losses during the race; it may take them more than a day fully to rehydrate.

Of course, more and more people around the world deal with heat not by sweating but by living in controlled (air-conditioned) environments, and extrasomatic solutions (shelters, furry clothes, fire) rather than biophysical adaptations have always been the preferred way of dealing with cold. Adjustments such as vasoconstriction (reducing the blood flow into skin and extremities) are good enough for feeling less cold during a desert night (Australian Aborigines are masters of this adaptation) but not for surviving in the Arctic. Many mammals are much better adapted for cold than humans but, in strict physical terms, we are unequalled as far as the conjoined attributes of bipedalism, outstanding heat disposal and endurance running are concerned. There is strong

HUMAN THERMOREGULATION (*cont.*)

evidence that endurance running, which originated in hominins about two million years ago, may have been a critical factor in the evolution of the human body. There is no doubt that without efficient sweating we could not be such great endurance runners or such a truly global species.

Foraging societies: gatherers, hunters, fishers

The energetic imperatives make it clear why (some notable exceptions aside) the typical diets of foraging societies were overwhelmingly vegetarian. As explained in the previous chapter, the amount of energy available to organisms feeding at higher trophic levels is reduced. As primary consumers – digging tubers, collecting seeds, gathering nuts and berries, and picking fruits – foragers had access to a food energy that was more than an order of magnitude more abundant than would have been available from consumption of animals feeding on that same phytomass. Moreover, not much energy is needed to harvest these foods; anthropological studies of foraging societies that survived into the twentieth century show that foraging for edible phytomass returned, even in arid environments, at least five times, commonly ten to fifteen times, and in some cases (particularly in harvesting tubers) more than thirty times as much edible energy as was expended in their gathering.

Of course, many of these harvests (particularly in temperate environments) were seasonal, and to maximize food availability most foragers exploited any edible phytomass: studies show that certain groups ate some part of scores of different plants (more than a hundred in some cases) although a few abundant species provided a large share of the overall intake. Energy-dense seeds and nuts (be it the acorns preferred by Native American tribes in California or the mongongo nuts preferred by the foragers in the Kalahari Desert) were obvious favorites. Their energy density ranges from 15 MJ/kg for grass seeds to just over 25 MJ/kg for piñon nuts, compared to less than 5 MJ/kg for tubers and squashes, and less than 1 MJ/kg for most of the edible leaves. Seeds and nuts have yet another key nutritional advantage, as they have relatively high protein content (many have more than twenty percent) while tubers, vegetables, and fruits are protein-poor (generally less than two percent).

In contrast to very (or at least relatively) abundant phyto-mass, animals, being one or more steps up the trophic pyramid, represented a much smaller amount of edible biomass, and ener-getic imperatives again explain both the most common and the most desirable choice of hunted species. There are many small and relatively abundant herbivores (particularly rodents) but these animals are also very agile, and once caught, yield only a small amount of meat. In tropical forests there are also many small arboreal herbivores, including monkeys and larger birds, but these species, dwelling high above the ground and camouflaged by dense canopies, are even more difficult to hunt. Studies of bow and arrow hunters in both African and Latin American rainfor-ests show very low rates of success, and hence very low energy returns, for the whole enterprise (or even a net energy loss when the hunters return empty-handed).

The best targets were the larger ungulates (tropical and subtropical antelopes, gazelles, boreal deer, and caribou), whose body mass (mostly between 50 and 500 kg) made it worthwhile to invest energy in an often protracted and demanding hunt. Moreover, all smaller ungulates are unaggressive, and large ungu-lates were often exceedingly abundant: before their mass slaugh-ter during the nineteenth century there were tens of millions of North American bison, and even now more than a million wildebeest annually migrate across the East African plains. Despite this, these herbivores were not easily caught or killed without weapons, or with only stones, bows and arrows, or spears, and the use of all these weapons required a close approach (difficult in an open grassland) and often just wounded the animal.

Perhaps the most remarkable method of hunting these fast ungulates was to run them down, that is, to pursue them for such a long distance that they got tired, slowed down, and could be killed or even caught alive. This method of hunting was documented (for chasing deer or antelopes) among the Paiute, Navajo, Goshute, Papago, Tarahumara, and Seri tribes in the North American southwest and northern Mexico, while the Kalahari Basarwa used it to run down duikers and gemsbok (particularly during the dry season), and some Australian Aborigine groups pursued kangaroos. The combination of the unique human abilities to run long distances at variable speed and to thermoregulate through sweating was what made these feats possible. Other hunters resorted to ingenious strategies to kill these animals, perhaps none more effective than the patient corralling of a bison herd into a confined space by enticing the herd with the imitated bleating of a lost calf, then using long drive lines made of stone cairns and hunters to channel them in the desired direction (with the help of young men camouflaged as wolves or coyotes) and their subsequent stampeding over a steep cliff.

HUNTING BISON AND MAMMOTHS

The evidence, an enormous accumulation of skeletal remains, at the most famous North American site of corral hunting – Head-Smashed-In in southern Alberta (since 1981 a world heritage site) – indicates that corralling had been used for millennia (this site had been used for more than 5,500 years).

Other large mammals had to be confronted one by one yet, as excavations of many prehistoric sites testify, hunters equipped only with spears were willing to take the risk of attacking woolly mammoth, an animal about one hundred times more massive than the largest man in the hunting group.

HUNTING BISON AND MAMMOTHS (*cont.*)

This preference for killing mammoths or bison has an energy-based explanation in the high value placed on the consumption of lipids, the nutrients with the highest energy density whose intake leaves an incomparable feeling of satiety. Hare, antelope, or even deer, animals that may have been easy to snare and certainly much less dangerous to kill, were almost pure protein: their meat had less than 6 MJ/kg and only a few percent of lipids, and so a small antelope or deer could yield only a few hundred grams of fat. In contrast, the bodies of large grazers, even if relatively lean, contained a much greater amount of lipids. A large male bison could yield more than 50 kg of fat, essential for making a highly energy-dense foodstuff. At sites such as Head-Smashed-In, the fat was rendered by throwing hot stones into water-filled pits lined with buffalo hides, the meat was dried in the sun and the bone marrow extracted. The dried meat was mixed and pounded together with the grease and marrow (berries were sometimes added) to make pemmican, a durable, high energy-density staple (and an obvious precursor of today's "energy bars").

A large woolly mammoth would yield ten times as much fat as a bison. This meant that the hunters not only obtained a large energy return (larger than most types of foraging) for their exertion and for the risk they took, but also that a large part of that return was in the form of extraordinarily filling fats. The hunting of megaherbivores required a co-operative approach at all stages of the enterprise: tracking, killing, and butchering required higher energy inputs through group participation but the effort returned much more energy than would have been ever accessible to a single hunter. There is no doubt that these benefits had an important role in the development of human group dynamics.

There was only one kind of hunting whose energy return surpassed that of killing terrestrial megaherbivores: fishing in the coastal waters regularly visited by enormous masses of migratory fish or whales. The Pacific Northwest, with its massive salmon runs and near-shore migration of baleen whales, offered the best opportunities of this kind, and the resulting energy surpluses allowed many tribes to do away with migratory foraging and set up semi-permanent or permanent settlements, some with substantial wooden structures. These groups reached the highest population densities for foraging societies, nearly one hundred people per square kilometer. In contrast, typical densities were less than ten people per square kilometer for forest dwellers: the large standing phytomass of those ecosystems is mostly locked in massive tree trunks and other inedible matter (climbers, leaves, shrubs) and the huge variety of tropical rainforest species means that individual trees or shrubs will be widely dispersed and their exploitation will require frequent change of campsites. Foragers in arid regions had average population densities another order of magnitude lower than the forest dwellers, at around one person per square kilometer.

Traditional agricultures: foundations and advances

Density comparisons show how even the least productive traditional agricultures increased the concentration and size of human settlements (Figure 12). Harvests in ancient Mesopotamia, along the Nile and on North China's plains could support around one hundred people per square kilometer of cultivated land; the highest averages of the pre-industrial era were about five times that. But one thing that was not substantially different was the average composition of diets. Much as in the case of foraging societies, energetic imperatives restricted the feeding choices of traditional field cultivators. Their diets were almost invariably dominated by the staple cereals and legumes, enriched by vegetables and fruits and marginally supplemented by animal foodstuffs (commonly including wild game). With low crop yields, it would have made no sense for those societies to feed their grains and tubers to meat-producing endothermic animals, because their metabolism would consume more

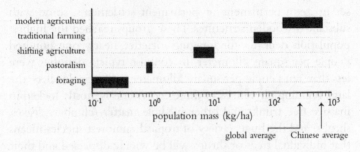

Figure 12 Population densities of different modes of food provision

than ninety percent of those energy inputs and so only a small fraction of the ingested grains and tubers would become available as meat. As shown in Figure 12, the more intensive ways of food production have been able to support more people per hectare of land, resulting in exponentially greater population densities.

Consequently, meat production was energized either by feeding the animals crop residues indigestible by humans or, in the case of ruminants, by grazing. Both these strategies had limitations. Crop residues (cereal straw and plant stalks were always most voluminous) had many other competing uses: in unforested regions they provided indispensable household fuel and animal bedding, they were used for thatching, as an excellent substrate for cultivating mushrooms, woven into many products (from beautifully finished Japanese *tatami* mats to ropes, hats, bags, and baskets), stuffed into pillows and horse collars, and pulped to make paper. And while grazing land was plentiful in some traditional farming societies, in others it was very limited, either by aridity (in large parts of the Mediterranean, in the Middle East), or high population density (in parts of China and on Java, virtually all suitable land was cultivated). In such places grazing was restricted to marginal lands (and so often only to sheep and goats), to road and canal banks, or harvested fields.

Traditional cultivators thus got most of their protein from the same plant foods that supplied their carbohydrates, supplementing it with a small (often irregular) intake of meat (but eating some, or even all, animals was proscribed in Buddhism, Hinduism, Jainism, Judaism, and Islam), dairy products (but many large populations, including the Chinese, Japanese, and the Natives of both Americas, had no milking animals), fish, and shellfish. Given the fact that most land was devoted to the cultivation of staple grains, and that animal food intakes were generally limited, traditional agricultural diets were also low in lipids, which supplied less than fifteen percent of food energy.

Unfortunately, many traditional diets were not just low in animal protein and lipids. All pre-industrial societies experienced recurrent food shortages that repeatedly reduced average energy intakes to below the levels compatible with healthy and active life (thus stunting of growth was commonplace), and that could even produce catastrophic famines. Europe's last famine was in Ireland, between 1845 and 1849, Asia's in China, between 1959 and 1961, when some thirty million people died (though the prime culprit here was not failed harvests caused by drought, but Maoist policies that catastrophically weakened the none-too-strong foundation of the country's still very traditional farming, which used virtually no inorganic fertilizers and no mechanization).

Nothing shows better the millennia-long stagnation or, at best, very slow improvements in average crop yields, than comparing the population that could be supported by one hectare of cultivated land. Reconstruction of Ancient Egyptian yields (a single crop grown with the help of the annual inundation by the Nile) shows an average density of just over one person per hectare under the Old Kingdom, rising to more than two by the beginning of the first millennium C.E. China's average reached that level only around 1000 C.E. (Sung dynasty); during the Qing dynasty it eventually peaked at nearly five people per hectare (with an almost completely vegetarian diet). Other small, intensively cultivated areas, such as Java and the Nile Delta, also achieved similar figures, but these were exceptions. The European average never rose above two people per hectare during the pre-industrial era and the United States' was even lower (albeit with higher meat production). By the end of the nineteenth century, average wheat yields were below 1 t/ha (or just 15 GJ of food energy) in the U.S., to about 1.3 t/ha in France; only the English and Dutch yields were around 2 t/ha.

The millennia of low yields, and hence uncertain food supply, had a number of causes. From the agronomic point of view, the cultivation of unimproved crop varieties (which channeled most of the photosynthate into inedible residue rather than harvestable edible parts), and the inadequate amount of nutrients (especially nitrogen) were particularly important. From the energetic point of view, the most important limiting factors were the inadequate power and relatively high energy cost of the only two kinds of prime movers available for field work; human and animal muscles. Human-powered cropping is best suited to intensive garden-type cultivation. Hoeing on a larger scale demands too much labor: in heavier soils it could take 200 hours to get one hectare ready for planting – too long to practice on an extensive scale. Draft animals cut that work dramatically: even a single ox, pulling an inefficient and heavy wooden plow, could do the task in about thirty hours, a pair of horses pulling a curved steel plow needed just three (in a lighter soil).

Bovines (cattle and water buffaloes) and equines (horses, ponies, mules, and donkeys) were the most important kinds of animals used for agricultural draft work (camels and yaks mattered only regionally). The body weight of these animals ranged from less than 200 kg for small donkeys to well over 700 kg for good horses and was the main determinant of their draft power, though this was also influenced by the animal's health and age, soil conditions (heavy clay soils were the most difficult to work), and by its harness. Typical performances were about ten percent of bovine weight (a draft of at least 30–40 kg) and fifteen percent of horse weight (60–100 kg). The sustained power of working animals ranged from no more than 300 W for smaller oxen to 700–800 W for good horses (you will recall that one horsepower equals 745 W). Healthy animals could carry on at this rate (with short rests) for hours.

The world has always had more working bovines than horses (Figure 13). Why this preference for less powerful and slower animals, that usually worked for less than ten years (compared to 15–20 for horses), were more difficult to train, and did not have the endurance of horses? Certainly, harnessing oxen is cheaper, they do not need metal shoes, and they are not as excitable – but the key reason is energetic. Ruminants have a highly specialized digestive system, and can (with the help of microbes that reside in their rumen) metabolize

Figure 13 A Chinese ox (left) and a French horse (right) harnessed to rotate wheels. Reproduced from *Tian gong kai wu* (1637) and *L'Encyclopedie* (1769–1772)

cellulose, which other mammals are either unable to digest or can use only poorly and in limited amounts. Ruminants do not compete for their feed either with other animals, or with people. Moreover, one bovine species, water buffalo, is a perfect fit for the tropics; able to walk well in muddy terrain, feed on aquatic plants, and metabolize its feed more efficiently than cattle. And all bovines eventually yield meat and leather, and their manure is a recyclable and welcome source of nutrients for the fields. That is why oxen (and even working cows), were always preferred by poor farmers, despite the fact that horses make far better draft animals.

Draft animals, whether rather weak or very powerful, provided very good energy returns. Assuming that people could work at 60–80 W, a strictly quantitative comparison makes even a small ox (energized solely by grass and crop residues indigestible by humans) about as powerful as five women. A strong horse consumed 4 kg of oats a day, and the cultivation of this grain needed land that would have yielded enough food grain (wheat or rye) for six adults – but that horse's work was equivalent to the day-long exertion of at least ten strong men. Such comparisons still miss the fact that these animals could do tasks that could not be accomplished even by large numbers of men, and that the animals could be harnessed in configurations large enough to perform work that would have been difficult, or impossible, to do any other way.

DRAFT HORSES

Some reasons for the superiority of horses as draft animals are obvious, others are subtle and peculiar. Many horse breeds are much heavier than oxen, and hence more powerful, and they also have greater endurance. But what may not be immediately apparent is that their fronts are heavier than their rears (by a ratio of 3:2), giving them a greater forward inertia. They also live longer than

DRAFT HORSES (cont.)

cattle, and, thanks to a unique attribute, they do not use any additional energy while standing (the suspensory ligaments and tendons lock their legs, so they do not have to tension their long leg muscles, as other animals do). For centuries, these advantages remained largely irrelevant. Even more efficient harnesses and iron horseshoes were not enough to make horses the dominant prime movers (in at least some regions): that shift took place only once crop yields increased enough to allow sufficient production of grain feed and a more widespread use of heavier, more powerful breeds. This is not to say that the two earlier advances were not important.

Collar harness (this originated in China but it took several centuries, until the end of the first millennium C.E., for its improved design to reach Europe) made the best use of an animal's powerful breast and shoulder muscles; iron horseshoes improved traction and prevented excessive wear of the horse's hooves. But horses can work at their full potential only when their roughage feed (grass) is generously supplemented by concentrates (cereal and leguminous grains). This became possible only as higher yields made more land available for planting feed crops, a development that began in Atlantic Europe in the late eighteenth century and reached its peak on the North American plains, in the late nineteenth and early twentieth century.

Heavier breeds (French Percherons, English Shires, German Rheinlanders) could deliver, steadily, more than one horsepower (up to 1 kW) and briefly more than 2 kW. In deep plowing, the main field task for which they were used, horse power was made more effective by the replacement of clumsy wooden plows (whose flat moldboards had high resistance and clogged easily), first with iron moldboards (in Europe during the seventeenth century) and then, starting in the middle of the nineteenth century, with smooth, curved steel plowshares that sliced through soil and turned it over with relatively little resistance. Only the combination of well-fed, powerful horses, and efficient steel plowshares made it possible, after 1860, to open the grasslands of North America, Australia, Argentina, and southern Russia for large-scale cultivation.

At least two good horses were needed to pull mechanical grain reapers and binders (introduced after 1831); large horse teams (many with more than two dozen animals) pulled the first

DRAFT HORSES (*cont.*)

grain combines, introduced in California in the late nineteenth century. This horse-based mechanization required large numbers of strong animals, and came at a high energy cost. In 1919, when the number of horses and mules working on the U.S. farms peaked, at just over twenty million, about twenty percent of the country's abundant farmland had to be devoted to their feeding. Needless to say, China or India, where all but a tiny portion of land had to be planted to food crops, could never repeat this achievement.

Biomass fuels: heat and light

The first urban civilizations – the lower Mesopotamian clay towns of Uruk, Ur, and Lagash (founded around 3200 B.C.E.) or the wooden cities of Shang China (around 2000 B.C.E.) – relied on the very same kind of fuels to cook their meals, fire their bricks, and smelt their metal as did the large, and (and at least in part) opulent, cities of the early modern world, such as seventeenth-century Venice, or eighteenth-century Paris. These societies, though continents and millennia apart, got their useful heat from converting the chemical energy in phytomass, mostly the wood of forest trees, but also from deliberately planted fuel-wood groves and wood charcoal. Where there were no nearby forests – on the alluvial plains, dominated by crop fields, in grasslands, in areas too arid to support more than small bushes – towns had to import fuelwood and charcoal, often from considerable distances, and peasants and poor city dwellers used any accessible phytomass.

Crop residues were used for fuel on all continents: cereal straw and the stalks of leguminous crops were usually the most abundant, but peasants also used cotton stalks, sugar cane leaves, and the roots of some crops. Where even these residues were in short supply people used the dried dung of animals, including cattle (on the Indian subcontinent and in Mongolia), yak (in Tibet), camel (in the deserts of Africa and Asia), and bison (on the North American plain it proved indispensable to the pioneering settlers of the nineteenth century). As already noted, wood's energy density depends on its moisture content, and that is why dead branches and fallen twigs were the preferred choice (also, their harvesting did not require any axes or saws). Air-dried wood has between 14 and 16 MJ/kg, compared to about 11 MJ/kg for air-dried straw, and 9–11 MJ/kg for air-dried cattle dung.

The relatively low energy density of phytomass fuels would be much less of a problem if they could be harvested with high power density. But, as we have seen, in the previous chapter, most forests store no more than 200 t of phytomass per hectare ($20 kg/m^2$) and so even if all of it could be harvested (and most of it is in large tree trunks that can be felled only by good metal tools), the yield would be more than $300 MJ/m^2$. Because a clear-cut forest may take between 50 and 100 years to return to its pre-harvest state, the power density of wood harvests must be calculated by dividing the energy total by the time it takes to regenerate the phytomass. Consequently, even if the harvest rate is divided by just fifty years, the actual sustainable power density of wood harvests would be merely $0.2 W/m^2$.

Charcoal, although an inherently better fuel than the wood from which it is made, gives even lower returns. Because it is virtually pure carbon, its energy density is about 30 MJ/kg, some sixty-five percent higher than air-dried wood; hence it makes an excellent metallurgical fuel, is easier to transport and store, and its largely smokeless combustion makes it an excellent fuel for unvented indoor cooking and heating in braziers. Traditional charcoaling was a very wasteful process, as only about twenty percent of the energy in the wood ended up in the charcoal, and so the power density of forest phytomass sustainably harvested for charcoal would be a mere $0.04 W/m^2$.

These rates must be contrasted with the typical thermal energy needs of a sizeable pre-industrial city. Depending on the mode of cooking (hardly any baking, common in Europe, was done in China, to save fuel), severity of climate (traditionally, no heating was done in China, south of the Yangtze, but Russian winters needed a great deal of wood), and on the amount of manufacturing that went on within the city walls (smithing, pottery, bricks), pre-industrial cities needed at least ten and up to thirty watts per square meter of their built-up area. This means that if they relied entirely on wood, they needed nearby areas of between 50 and 150 times their size in order to have a sustainable phytomass, and from 85 to 250 larger if the fuel supply were divided between wood and charcoal. This alone would have precluded a megacity (ten million people) in any pre-industrial society whose thermal energy came from phytomass.

Things got even worse as charcoal-based iron smelting began to expand from a small-scale activity to mass production. Although rates improved with time (by the mid-eighteenth century a new blast furnace consumed less than one-tenth the charcoal per unit of hot metal output than its medieval predecessor), the availability of forest phytomass rapidly became the key factor in determining the future expansion of iron smelting. I calculate that, in 1810, America's iron industry claimed annually $2500 \, km^2$ of forest, an area easily accommodated by the country's rich forest resources. A century later the need would have been nearly $170,000 \, km^2$ of forest (twice the size of Austria); this amount of carbon could only be delivered by moving from charcoal to coke, made from coal. In the U.K., this transition was largely accomplished a century before.

Harvests of crop residues gave even lower power densities, especially in those temperate areas where only a single annual crop could be grown. With a typical pre-modern yield of 1 t/ha of cereal grains and a straw:grain ratio of 2:1, there would be less than $0.1 \, W/m^2$ in the residual cellulosic phytomass. But competing uses (bedding, feeding, thatching, manufactory) would leave only a part of this harvest for fuel. Even assuming that half could be collected for fuel, we would get no more than a tonne per hectare or less than $0.05 \, W/m^2$. A peasant family, living in a small one-room house in a temperate climate, could meet a large part of its cooking and heating needs by burning straw, but a sizeable city could not be energized by crop residues, because they would have to be collected from an area up to 600 times larger than the city.

And phytomass fuels are a source of indoor air pollution. They can generate high levels of poisonous carbon monoxide, while poorly vented combustion, in shallow pits or fireplaces, produces high concentrations of fine particulates, including various carcinogens. Repeated inhalation of this smoke leads to impaired lung function and chronic respiratory diseases (bronchitis, emphysema). These impacts (still affecting millions of people in poor countries, where the inefficient combustion of phytomass remains the primary source of cooking heat) can only be reduced or eliminated by more efficient stoves with grates and pipes connected to chimneys to provide proper venting of combustion gases.

Lighting progressed slowly, from open fireplaces and resinous torches, to clay lamps fueled first with animal fat (the oldest go back some 40,000 years) and then (much later, in agricultural societies) with various plant oils. Candles (made of beeswax or beef tallow) appeared after 800 B.C.E. and better illumination was possible only by increasing their numbers. As well as their obvious inconvenience (limited durability, smoky combustion, fire risk), the efficiencies of converting the chemical energy of candle fats to light were pitiful, always less than 0.01 percent. Even the best oil lamps, of the late eighteenth and early nineteenth century, had efficiencies no better than 0.03 percent. These had glass chimneys and regulated wicks, and were fueled, until the 1860s, by oil rendered from the huge carcasses of sperm whales, slaughtered around the world by large whaling fleets. Although the earliest matches were used in China during the sixth century C.E., they made it into Europe only after 1500 and the modern (safety) variety was sold only in 1844. The invention of gas distillation from coal, and the introduction of kerosene (refined from crude oil, on a large scale only after 1870), the invention of the first practical incandescent light bulbs (1879–1880), and the central-ized, large-scale, generation of electricity (1882), made house-hold, street, and industrial lighting much more convenient and eventually (with the introduction of fluorescent and sodium lights) orders of magnitude more efficient.

Pre-industrial cities: transport and manufacturing

The emergence of cities marked the beginning of sedentary societies that produced enough surplus food energy to allow a portion of the population to engage in activities other than crop cultivation and animal husbandry. Cities were thus the primary drivers of increasing social and economic complexity, including the emergence of religions, symbolic architecture, diverse visual arts, writing systems and literature, laws and rules of economic conduct, and the empirical observations and studies of natural phenomena that gradually evolved into science. For millennia, these intellectual pursuits involved only a small share of the total population, as the small agricultural surplus and the strictures of phytomass energy supply did not allow the urbanized population to rise, at best, much above ten percent of the total.

However, even in antiquity and the Middle Ages there were some large cities. Rome had, at the end of the first century C.E., more than half a million people, Harun ar-Rashīd's (caliph of the *Thousand and One Nights*) Baghdad, had, in the early ninth century

C.E., 700,000, and Changan, its great Asian contemporary and capital of the Tang dynasty, peaked at about 800,000. But the total number of such large cities remained low until the beginning of the nineteenth century. By 1800, Beijing, the capital of the Qing dynasty, surpassed one million, and London was nearly as populous, but both in China and Europe, urban populations were still only about ten percent of the totals. For inland cities not situated on the banks of navigable rivers, the limits due to food and fuel provision were compounded by inefficient land transport, which greatly restricted the economically viable import of daily necessities.

TRADITIONAL LAND TRANSPORT

Before the introduction of railroads, all land transport relied on the same prime movers as traditional farming: on human and animal muscle. The animals that dominated field tasks, oxen and horses, were also the most important draft beasts, used for transporting goods and some heavy construction tasks. Camels were load carriers throughout North Africa, the Middle East, and parts of Asia (Beijing received caravans into the 1940s), elephants in the Indian subcontinent and Southeast Asia (especially in forest tasks), yaks in Tibet, and llamas in the Andean countries of Latin America. In mountainous regions, porters (carrying 25–40 kg) and small pack animals remained important throughout the twentieth century.

Pre-industrial wheeled transport evolved very slowly, and in many places actually regressed for centuries. This was most notable in Europe, where the ancient Roman network of excellent *viae* (surfaced with gravel concrete, cobblestones, or flagstones) was only equaled, after long medieval and early modern decline, during the nineteenth century. Road surface and vehicle design are the key determinants of friction and hence of the energy needed for wheeled transport. On smooth and hard surfaces, only about 30 kg is needed to move one tonne, on loose or rough surfaces the force must be four to five times greater, in sand and mud up to ten times. Heavy wagons and the lack of lubrication aggravated the problem: lighter designs, lubricated axles and (by the seventeenth century) ball bearings lowered the overall friction.

TRADITIONAL LAND TRANSPORT (*cont.*)

Poor roads, weak animals, and inefficient vehicle designs limited the maximum loads and typical speeds of goods transport. The Roman limits for ox-drawn wagons were 490 kg and 15–20 km a day (messengers riding, and changing, fast horses could do more than 300 km). That is why, after Egyptian grain arrived at Ostia (the ancient Roman port on the Tyrrhenian Sea), it was reloaded on to barges rather than wagons for the trip, of less than 25 km, to the city. Only by the eighteenth century did better wagon designs and surfaced roads combine to raise loads (to more than one tonne) appreciably, and to increase the speed of transfer. After 1840, long-distance horse-drawn transport of goods was rapidly replaced by rail, but draft animals remained indispensable for distributing goods and moving people within cities until the first decades of the twentieth century (at Queen Victoria's death, in 1901, London had about 300,000 horses), when they were displaced by electric and internal combustion engines. The stabling and feeding of horses (and the removal of their voluminous waste) were a challenge to the design and management of all large, pre-World War I, cities.

Wind-powered water-borne transport was always cheaper and faster, but its full potential could not be realized as long as there were no efficient sails or maneuverable ships. The large, oared ships, used in the Mediterranean from antiquity until the seventeenth century, were inherently inefficient and were overwhelmingly used for military campaigns, rather than carrying goods. In basic physical terms, sails are fabric airfoils that should maximize lift and minimize drag. Everywhere, their initial form was the simple, inefficient square or rectangle, seen in Egyptian tomb paintings or on Greek pottery. More than two thousand years ago, the Chinese adopted the characteristic batten-strengthened lug sail (familiar from images of Chinese junks), and, by the seventh century C.E., came the triangular sails of the Arab world and Indian Ocean.

For centuries Europe was a follower rather than an innovator in ocean shipping: the stern-post rudder and magnetic compass were among key imports (both from China). Only by the late Middle Ages did European ships gain the ability to sail close to the wind, by combining square and triangular sails, and so (equipped with heavy guns) became the prime tools for projecting European colonial power around the world. This began before 1450, with the Portuguese voyages along the coast of Africa, and ended only during the late nineteenth century with the continent's final partitioning between Great Britain, France, Spain, Portugal, Belgium, and Germany. The first crossings of the Atlantic (Columbus, 1492) and the Pacific (Magellan, 1519) were made by vessels that were the same size as standard Roman cargo ships (100–200 t); three centuries later the ships were an order of magnitude larger, and also considerably quicker. The famous China clippers averaged as much as 5 m/s on long journeys, twice as fast as the fastest Roman vessels (Figure 14).

The movement of food, fuel, construction materials, and a limited range of consumer goods to pre-industrial cities was

Figure 14 A nineteenth-century clipper

energized largely by draft animals and wind, but the construction of buildings, roads, bridges, and aqueducts, and the variety of artisanal manufacturing done in the cities relied overwhelmingly on human labor whose effectiveness was enhanced by simple mechanical devices, based on the fundamental principles of lever, inclined plane and pulley with elaborations and combinations such as wooden wedges, screws, wheels, windlasses, treadwheels, and gearwheels. These simple tools were sufficient to complete such remarkable structures as the megaliths of Atlantic Europe, the giant heads of Easter Island, the stepped pyramids and temples of Mesoamerica and, most impressively, the unequalled and massive stone pyramids of Egypt's Old Kingdom. The technical details of building these remarkable projects still remain unclear and may be never fully understood. Eventually, these simple machines evolved into very sophisticated designs capable of the remarkable feats of Renaissance and early modern engineering. They were needed because of the problems inherent in simply multiplying the number of laborers. A supervisor at an ancient construction site could command hundreds of people – but it would take fewer than a dozen to completely encircle a heavy stone and so their combined power would be too small to lift it; on the other hand, the same dozen could move it with the help of levers, inclined planes, or ropes and pulleys.

But cities needed unprecedented amounts of processed food (wheat had to be ground to make flour, or seeds pressed to make oil), and metal items (iron, copper, and lead had to be smelted, forged, and shaped into the final products) and neither human nor animal muscles were powerful enough, even when their numbers were multiplied to the maximum, to meet these needs. That is why most of these tasks were the greatest beneficiaries of the first non-animate prime movers devised by humans: waterwheels and, some centuries later, windmills.

The uncertain beginnings of waterwheels date to the first century B.C.E., but those simple wooden machines with vertical axes made little difference either in Classical Greece or Imperial Rome: slave labor was abundant, so neither of these civilizations developed any large-scale centralized manufacturing. Even after the unraveling of the Western Roman Empire it took a long time for typical waterwheels to reach capacities surpassing the power of large, harnessed, animal teams. Similarly, the origins of windmills are also unclear, but inefficient vertical-shaft machines were working in some areas of the Middle East by the tenth century C.E. Eventually, both waterwheels and windmills were in fairly common use in many Muslim countries, but they were particularly embraced in late medieval and Renaissance Europe. The tasks done by the rotary movement delivered by these prime movers ranged from polishing tiles to pumping water, and from powering blast furnace bellows to forging heavy iron pieces. These advances and adaptations were critical in setting the stage for the emergence of machine-dominated Western civilization.

The early modern world: the rise of machines

The process of gradual mechanization extended over several centuries. Nothing illustrates better its accelerating progress during the early modern era than a comparison of the excellent woodcuts included in Agostino Ramelli's (1527–1608) sixteenth-century classic *Le Diverse et Artificiose Machine* (first published in 1588) and the elegant engravings that appeared in the world's first richly illustrated encyclopedia, published under the editorship of Denis Diderot (1713–1784) in France (the last of 28 volumes of *L'Encyclopedie* was completed in 1771). Ramelli's illustrations of commonly used machines (waterwheels, pumps, windmills), tools (gearing, chains, cams), and his own new designs show devices that are ingenious but made overwhelmingly from wood, and are obviously heavy and clearly not very efficient. In contrast, L'Encyclopedie is full of illustrations that convey not only a greater sophistication of design of machinery but also a much higher exactitude of execution. A century later both of these qualities were even more apparent, as the two old prime movers, water-wheels and windmills, were now providing energy for Western industrialization.

Waterwheels were the most advanced energy converters of the early modern world. Their use spread to scores of highly special-ized tasks, that neither groups of men nor harnessed animals could perform. But their power remained limited (in Europe typically no more than 4 kW by the end of the seventeenth century), and their rough wooden gearing and inadequate lubrication resulted in low conversion efficiencies. Matters had improved during the eighteenth century, and there is no doubt that early Western industrialization relied on water power as much as it did on steam engines. Even in 1849, in the U.S., the installed power of all steam engines was only twice that of all waterwheels. In some regions (New England, parts of France, Scotland), and in some industries (particularly textiles and ore mining) waterwheels and their more efficient successors, water turbines, remained the dominant prime movers into the latter half of the nineteenth century.

Waterwheels

Every possible design was used, but two kinds of vertical wheels (mounted on horizontal axes and transmitting their power through right-angled gearing) were most common. Undershot wheels (rotating counter-clockwise) were driven directly by natural water flow (the faster the better, as the wheel's power increased with the cube of water speed) or, more commonly, by water flowing through a flume (Figure 15). The most efficient undershot wheels were Poncelet wheels, whose curved blades converted at least twenty percent of the water's kinetic energy into rotary motion. Overshot wheels (they could rotate either way) were driven primarily by the impact of descending water channeled through a flume and hence they could be placed on slow-flowing streams. Their efficiencies were commonly above fifty and could be up to eighty percent. There were also breast wheels, driven by water impacting just below or just above the elevation of the center shaft, and even many small horizontal wheels, of ancient provenance.

The power of waterwheels increased both through the massing of smaller units and construction of machines of unprecedented capacity. The world's largest water-driven machine (completed in 1854 by Great Laxey Mining Company on the Isle of Man) showed both complexity and sophistication of design. Streams above the wheel were guided into collecting tanks, the water then piped to a large masonry tower, and let fall through a flume on the top of the wheel to rotate it counter-clockwise (the pitchback over-shot design). The maximum peak power was

Figure 15 Late eighteenth-century French undershot waterwheel (reproduced from *L'Encyclopedie*)

over 400 kW and in its everyday pumping operations the wheel generated about 200 kW. The efficiencies of wheel operation improved greatly once iron began replacing wooden hubs and shafts but the first all-iron, smoothly geared, and well-lubricated wheels appeared only after 1800.

The metal waterwheels of the first half of the nineteenth century led directly to the first water turbines. Benoit Fourneyron (1802–1867) completed the first (with radial outward flow) in 1832 to power forge hammers, and soon afterwards had a better design to run a spinning mill. A much more successful design was patented in the U.S. in 1838 by Samuel B. Howd (1807–1881) and improved and commercially deployed by James B. Francis (1815–1892). During the 1880s came Lester Allen Pelton's (1829–1908) impulse turbine, driven by water jets discharged into peripheral buckets (a design that is best suited for very high water heads), while in 1913 Viktor Kaplan (1876–1934) patented a reaction turbine with adjustable vertical-flow propellers, used for low water-head projects. All early turbines were used to produce rotary power for industrial applications but after the 1880s turbines were also used to generate electricity.

The greater sophistication and higher efficiency of waterwheels was mirrored by windmill design (Figure 16). Around 1600, Dutch millers introduced a canted leading edge to their large, flat (and drag-inducing) blades, an innovation that improved lift and reduced drag. Then, cast metal gearings began replacing wooden assemblies, and after 1745 English millers began using fantails, which powered a winding gear that turned the sails automatically into the wind and so did away with the laborious task of turning the cap manually. English millers were also, by the end of the nineteenth century, the first users of true airfoils and aerodynamically contoured blades (much like modern airplane propellers). Automatic regulators, smooth transmissions, and the low prices brought by large-scale production made smaller

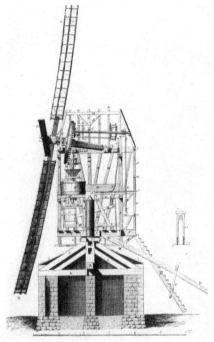

Figure 16 Section through an eighteenth-century French windmill (reproduced from *L'Encyclopedie*)

wind machines affordable and they had several important roles in opening up America's Great Plains (Figure 17).

These small converters (their typical rated capacities were as low as 30W and mostly less than 500W) did not have a few large sails but rather many narrow blades (slats) radiating from solid or sectional wheels equipped with governors and rudders and were placed atop wooden towers. These wind machines were widely used for pumping the water needed by the household and domestic animals, and for the steam locomotives crossing the continent. Several million of these small machines were sold

Figure 17 A late nineteenth-century windmill on the U.S. Great Plains Plains (reproduced from Wolff, A. R. 1900. *The Windmill as Prime Mover*)

in the U.S. and Canada during the second half of the nineteenth century, many more than the total number of windmills operating in Europe at that time. But European wind machines (most of them, about 30,000 in 1900, in the countries bordering the North Sea) were more powerful (surpassing 3 kW on average).

4

Energy in the modern world: fossil-fueled civilization

No terrestrial civilization can sever its dependence on solar energy; its incessant flow sustains the biosphere and powers the photosynthetic production of our food. Every civilization is thus, fundamentally, solar but the modern world has modified this in two important ways: it depends on fossilized stores of solar energy extracted as coals and hydrocarbons (crude oils and natural gases), and it increasingly relies on electricity, which it generates by burning fossil fuels, harnessing solar radiation (mostly indirectly as water and wind), using the Earth's heat (geothermal energy), and by nuclear energy (Figure 18). Traditional societies either drew their food, feed, heat, and mechanical power from sources that were almost immediate transformations of solar radiation (flowing water and wind), or harnessed it in the form of biomass and metabolic conversions that took a few months (crops harvested for food and fuel), a few years (draft animals, human muscles, shrubs, young trees) or a few decades (mature trees) to become usable.

In contrast, fossil fuels were formed through slow but profound changes of accumulated biomass under pressure and heat, and their ages (with the exception of some relatively young peats) range from millions to hundreds of millions of years old. A useful analogy is to see traditional societies as relying on instantaneous (or minimally delayed) and constantly replenished solar income, while modern civilization is withdrawing accumulated solar capital at rates that will exhaust it in a tiny fraction of the time that was needed to create it. Traditional societies were, at least in theory, energetically sustainable on a civilizational timescale of many millennia. But, in practice, many caused excessive deforestation and soil erosion, and overtaxed their labor. In contrast, modern civilization rests on unsustainable harnessing of the solar inheritance, which cannot be replenished on a civilizational timescale (so much for all that talk about enhancing the sustainability of modern economies: until we adopt an entirely new energy foundation it remains politically correct but scientifically risible).

But this very dependence on fossil fuels has given us access to energy resources that, unlike solar radiation, are both highly concentrated and easy to store. As a result, both aggregate and per caput energy consumption of modern societies have risen to unprecedented levels, and the rises are even higher when historical comparisons are made in terms of actually delivered energy services (heat, light, motion) rather than in terms of primary energy use. This is due to continuing technical advances, which have improved the typical efficiencies of nearly all the principal energy conversions. The affordable abundance of more efficiently used fossil energies has transformed every productive sector of the modern economy: field machines and agrochemicals have displaced animate labor and organic recycling in farming, mechanization has eliminated heavy exertion in extraction of mineral resources and has ushered in the era of mass production in manufacturing, new prime movers have revolutionized transport, and new forms of communication and information processing have had an even more profound effect on the service sector, which now creates most new wealth in affluent countries.

In turn, these advances changed the structure and dynamics of modern industrial and post-industrial societies. First and foremost, they raised average food availability far above subsistence

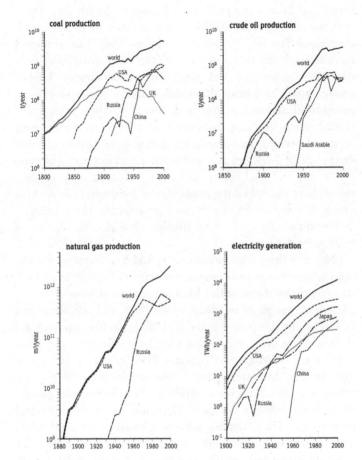

Figure 18 Global coal, crude oil, and natural gas production, and electricity generation during the twentieth century (plotted from data in UN and BP statistics)

needs, and assured the provision of adequate health care. These were the two key factors behind the steady increase in average life expectancy. Second, they produced the worldwide trend toward increasing urbanization, and the cities that have provided unprecedented occupational and intellectual opportunities. Higher average wages have made the middle class the dominant income group in all affluent countries and allowed it to buy an unprecedented variety of goods and services. Third, new techniques have increased personal mobility to levels that were hard to imagine even just half a century ago, and transformed international trade into a mass-scale and truly global affair. Finally, advances in electronics have multiplied the availability of information to such an extent that any owner of a PC or a smart phone can – instantly and inexpensively – access information equivalent to a library of millions of volumes.

None of these transformations would have happened without the conversion of an increasingly higher share of fossil fuels into electricity, the most flexible and convenient form of energy. Electricity has given us inexpensive, clean, and reliable means to extend the day, power individual machines (from tiny dental drills to giant earth excavators) with high efficiency and accuracy, propel trains at speeds greater than 300 km/h, and create entirely new industries (from metal smelting in electric arc furnaces to electrochemical syntheses). Electricity also powers the still-expanding universe of electronic communication and information. These systems were first based on vacuum tubes, after 1950 on increasingly more complex transistors, after 1960 on integrated circuits, and finally, starting in 1971 with Intel's 4004, on microprocessors, whose performance has been doubling every two years (conforming to Gordon Moore's famous law).

On the negative side, the technical advances made by high-energy fossil-fueled civilization have accentuated the gap between have and have-not nations, with access to information (through ownership of electronic devices, or purchases of printed matter) and opportunities to enjoy a high-quality life (rising life expectancy, choice of occupation) becoming even more unequal than average incomes. Extraordinary energy investments have made it possible to develop nuclear weapons (and hence to contemplate the nearly instantaneous annihilation of much of humanity, as a result of a thermonuclear exchange between superpowers) and resulted in an unprecedented degradation of the global environment.

The threat of rapid global warming has, during the twenty-first century, emerged as the most intractable consequence of the world's dependence on fossil fuels. The extent of this change remains highly uncertain, but the latest scenarios by the Intergovernmental Panel on Climate Change put the most likely increase of average global temperature at between +1.8°C (a manageable rise) and +4°C by 2100 (clearly a very worrisome change, especially when the upper range of that scenario is as high as 6.4°C). Regardless of the outcome of this great challenge, the longevity of fossil-fueled civilization cannot match its traditional predecessors. Its end will come long before we exhaust the finite stores of fossil fuels, as their higher extraction costs will lead to a greater reliance on solar energy flows converted by new techniques, or to development of new non-fossil sources of energy.

Coal: the first fossil fuel

Coals (the plural is more accurate, because of considerable differences in the composition and quality of these solid fuels) are sedimentary rocks, dominated by combustible organic material and containing varying amounts of incombustible mineral matter and water. All coals originated as plants, deposited in swampy environments, partially decomposed, covered by other sediments and subjected to high pressures and temperatures for up to 350 million years. Many of these plant species are still around, albeit in drastically smaller forms: next time you see foot-tall horsetails and thumb-tall club mosses growing in wet sylvan spots, imagine them as ten to thirty meter tall trees in the swampy forests inhabited by the first reptiles and large winged insects. More recently (ten to twenty million years ago) the buried trunks of ashes, maples, and willows began to be transformed into younger coals.

COAL TYPES AND COMPOSITION

Differences in the initial plant composition and the subsequent metamorphic processes explain the wide variation in quality apparent not only between coals from different fields but also between coals from a single field or even a single seam. The best coals, the jet-black anthracites, have their origin in the Carboniferous period (354–290 million years ago), as do the good-quality bituminous (hard, black) coals. Some lignites (Europeans call them brown coals), covered by only shallow sediments (and hence not subject to high pressures and temperatures) date from the Carboniferous era, but most are much younger, as they were produced by the transformation of plant material during the Tertiary period (65 million years ago). The poorest lignites (lighter colored and crumbly) have an energy density lower than wood, because most of their mass is moisture and ash. In elemental composition, the poorest lignites are less than fifty percent carbon, anthracites more than ninety percent, and bituminous coals mostly between seventy to seventy-five percent. This means that the most commonly used bituminous coals have energy densities about fifty percent higher than air-dried wood.

Consequently, energy stored as coal will occupy less space, require less frequent stoking of stoves and furnaces, and untended fires will last longer. On the other hand, the costly and dangerous underground mining is the most obvious disadvantage of coal extraction, while the presence of relatively large volumes of ash (bituminous coals typically contain about ten percent of incombustibles, mostly oxides of iron, silica, and alkaline elements) and sulfur are its greatest environmental drawbacks. Burning coals in lumps on grates (as is common in stoves and small furnaces)

COAL TYPES AND COMPOSITION (cont.)

produces bottom ash that has to be removed, while the combustion of finely milled coal (with particles as fine as flour) in large power plant and factory boilers generates fly ash that, if not captured by electrostatic precipitators, would slowly cover the surrounding and downwind areas with deposited dirt.

Sulfur is present in wood as a mere trace, but makes up, typically, about two percent of bituminous coals. Some is organic sulfur that was in the ancient plant proteins, concentrated by the prolonged pressure and heat, but sulfate sulfur and pyritic sulfur are often present in bituminous coal. The last kind is found as large shiny crystals (fools' gold) embedded in the black matrix. No matter what its origin, the combustion of sulfur generates sulfur dioxide (SO_2), a highly reactive gas, readily oxidized by atmospheric reactions into sulfates, the principal contributors to acid rain. Given the high carbon content of most coals it is also not surprising that the combustion of these solid fuels generates more carbon dioxide per unit of released energy (that is it contributes more to anthropogenic emissions of greenhouse gases) than any other fossil fuel.

Because some coal seams outcrop at the surface, or are covered by very shallow sediments, coal has been known and used since antiquity, locally and on a small scale, for heating, pottery making, and in metallurgy, from the Han dynasty in China (2000 years ago) to the Hopis of Northern Arizona (800 years ago). The oldest European records of industrial coal combustion are from the beginning of the twelfth century, in Belgium, and the early thirteenth in England. But the epochal transition from biomass fuels got underway only after wood resources were seriously depleted. This first took place in England, and was due not only to the rising demand for charcoal (needed, above all, for iron-making in blast furnaces) but also because of the growing need for timber for housing and shipbuilding. By Cromwell's time (the mid-seventeenth century), almost all the coalfields that later fueled English industrialization were already in operation, and by 1700 the country was producing three million tonnes of coal a year.

A century later that rose to 10 million tonnes, as coal's uses expanded from being a source of direct heat to being the feedstock for producing coke, and fueling the newly invented steam engines. In contrast to coal combustion (where flame temperatures are as high as 1650°C), coal pyrolysis involves gradual heating, up to about 1100°C. This thermal decomposition in the absence of oxygen produces gases, liquids, and coke – a highly porous but strong form of carbon, which eventually replaced charcoal as the fuel in blast furnaces. Coking was pioneered by Abraham Darby (1678–1717) in 1709, but widely adopted only after 1750. Coke's availability severed the dependence of iron making from charcoal, its ability to support heavier charges of ore and limestone opened the way for larger blast furnaces, and the higher temperature of coke-smelted iron made for better metal castings.

Coal's consumption really took off with the introduction of efficient steam engines, the first machines designed to convert the chemical energy in fuels to mechanical energy. The first commercial engines were designed by Thomas Newcomen (1663–1729) during the first decade of the eighteenth century, but because they condensed the steam on the underside of the piston (cooling it with every stroke) they were very inefficient, converting only about 0.5 percent of the coal's chemical energy into reciprocating motion. James Watt's (1736–1819) famous improvements, patented in 1769, included a separate steam condenser, an insulated steam jacket around the cylinder, and an air pump to maintain the vacuum (Figure 19). Watt also designed a double-acting engine (with the piston also driving on the down stroke) and a centrifugal governor to maintain constant speed with varying loads.

The typical power of Watt's engines was about twenty-five horsepower (about 20 kW); the largest engines, built in partnership with Matthew Boulton, were five times more powerful,

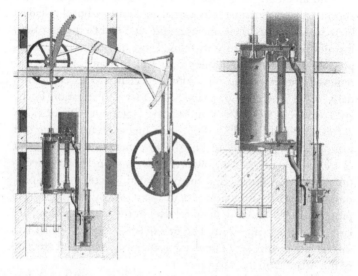

Figure 19 James Watt's steam engine (left) and detail of separate condenser (right) (reproduced from Farey, J. 1827. *A Treatise on Steam Engines*)

matching the performance of the largest contemporary water-wheels. These more (up to five percent) efficient engines were not only used in coal mines to pump water and operate winding and ventilating machinery, but were also gradually adopted by a growing number of industries whose locations were previously restricted by the availability of flowing water or steady winds. Iron making was a notable beneficiary, where steam engines were used to operate blast furnace bellows. In manufacturing, a single engine in a large workshop or a factory often powered a rotating axle from which a number of belts transmitted power to individual (weaving, grinding, boring, polishing, etc.) machines.

The inherently large size of steam engines was not a major consideration in stationary industrial applications, but this had to be reduced (that is, their operating pressure had to be increased) for mobile use. Because Watt refused to experiment with high-pressure engines, their development had to wait until after his renewed patent expired in 1800: the celebrated inventor thus delayed the machine's progress. Soon after its expiration Richard Trevithick (1771–1833) in England (1804) and Oliver Evans (1755–1819) in the U.S. (1805) built high-pressure boilers that were first tested on steam boats. The concurrent development of steamships and steam-powered railways soon led to successful commercial uses and, a mere generation later, to an enormous extension of these modes of transport. By the 1830s, the proven paddle-wheel designs of river boats were replicated in larger ocean-crossing ships. The first reliable screw propeller was introduced in 1838, the same year as the first westbound crossing of the Atlantic under steam power.

Large steamships then rapidly replaced unreliable sailing ships, first on the North Atlantic route (where they eventually cut travel time from more than two weeks to less than six days) and then on other inter-continental runs. Larger engines and, after 1877, the use of steel in hull construction led to luxurious passenger liners. Steamships carried most of the fifty million emigrants who left Europe between 1850 and 1914, while steam-powered naval vessels provided new means for the projection of Europe's colonial power. Steam engines transformed land transport in a similarly rapid and radical fashion. The first public railway (from Liverpool to Manchester) opened in 1830 but its first locomotive – George Stephenson's (1781–1848) famous *Rocket* – soon appeared laughably slow. By 1850, the fastest locomotives travelled at over 100 km/h; the relentless extension of railways soon spanned Europe and North America (but the completion of the trans-Siberian railway had to wait until 1904). A profusion of new locomotive designs brought ever more efficient and faster machines. By the end of the nineteenth century, speeds over 100 km/h were common and locomotive engines had efficiencies of more than twelve percent.

Besides providing coke, heat, and stationary and mobile power, coal also became a leading source of urban light, as its gasification produced low-energy coal, or town, gas. This first non-biomass source of light was introduced in 1805 in English cotton mills and London got a household supply in 1812. Coal gas was a very inefficient source of light: its combustion converted no more than 0.05 percent of the coal's energy into visible radiation, which meant that it also generated plenty of heat, water vapor, and carbonic acid. But gas lamps dominated urban lighting, indoors and out, until the 1880s, when they began to be replaced by incandescent lights.

Many technical advances were needed to meet the rising coal demand during the nineteenth century but what changed little was the reliance on heavy, and dangerous, labor in underground coal mining. Horses were used for hauling coal to the bottom of a pit, steam engines powered hoists and ventilators, but all other tasks were done by men, women, and children. The fuel was extracted from seams by miners, who often first had to walk several kilometers to reach their work faces before spending hours crouching or lying as they wielded their picks and mallets in confined spaces where they breathed coal and rock dust and were constantly endangered by cave-ins from unsupported ceilings and high methane levels that caused recurrent, and often deadly, explosions. Moving the fuel from the work faces to the loading points was often done by women and young girls, and young boys were employed throughout mines for many lighter duties. Perhaps no portrayal of these almost unbearably taxing and dangerous conditions will ever surpass Émile Zola's (1840–1902) *Germinal*, a painfully faithful description of conditions in the northern French mines of the late 1860s that was repeated, with slight variations, in all the coal-mining nations of the nineteenth century.

The world's use of coal eventually surpassed that of wood and crop residues: it is most likely that the scale tipped sometime during the 1890s. England entered the nineteenth century as a coal-dominated economy (in 1800 it produced eighty percent of the world's coal, only losing its leading place, to the U.S., during the 1870s). Most of Western and Central Europe accomplished the transition before 1870, while the U.S. derived more energy from coal than from wood until the early 1880s, and in Russia wood dominated until after World War I. Coal fueled the transition of traditional artisanal economies to modern mass manufacturing and made steam engines the most essential prime movers of the industrial revolution. Yet they remained relatively massive and inefficient machines, with a limited power capacity: the first making it a poor choice for fast road vehicles, and the second an awkward choice for increasingly large electricity-generating plants. The first drawback was countered by the introduction of internal combustion engines, and the second by Charles Parsons's (1854–1931) invention, the steam turbine.

PARSONS'S STEAM TURBINE

Parsons's steam turbine was the first important machine not to be invented on the basis of practical tinkering but to be designed *de novo* because thermodynamic principles indicated that such a machine was possible. Parsons was not the first engineer to design a steam turbine: Carl Gustaf Patrick de Laval (1845–1913), best known for his centrifugal cream separator (until its introduction all butter had to be churned manually), revealed his impulse steam turbine in 1882. But Laval's concept could not be easily converted into a practical machine, as his turbine blew the steam from trumpet-shaped nozzles on to the angled blades of a rotor and this impulse resulted in rapid rotation rates (in excess of 40,000 revolutions per minute (rpm)) and large centrifugal forces, which could not be withstood by any materials available at that time.

PARSONS'S STEAM TURBINE (cont.)

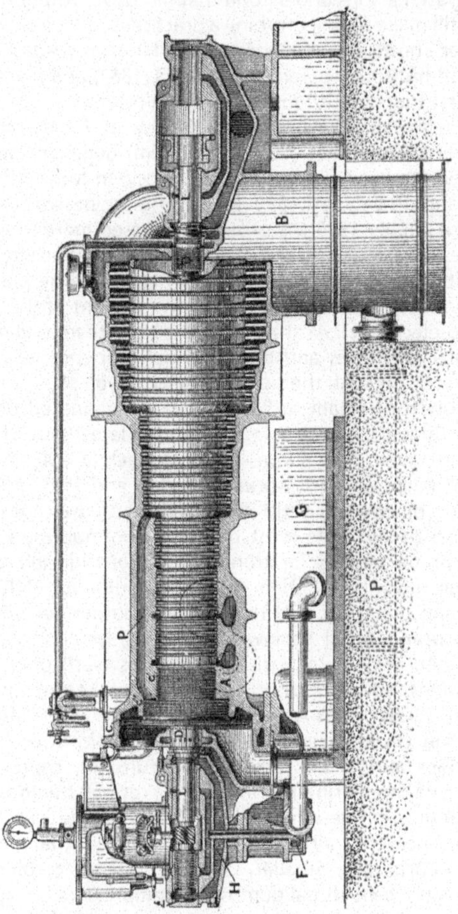

Figure 20 Section through Parsons's 1 MW steam turbine. Reproduced from the 1911 (eleventh) edition of *Encyclopedia Britannica*.

PARSONS'S STEAM TURBINE (*cont.*)

In contrast, Parsons understood that moderate rotation velocities could still make a steam turbine a practical and widely adopted prime mover. The development of steam turbines proceeded rapidly: Parsons filed his British patent on April 23, 1884, and a year later built the first tiny (7.5 kW) machine, which rotated at 18,000 rpm but whose efficiency was unacceptably low, at 1.6 percent. The first commercial machines (75 kW, 4,800 rpm) began to generate electricity (with about five percent efficiency) in Newcastle upon Tyne, England, in January 1890. A 1-MW machine was ready by 1900 (Figure 20). By 1910 the capacity increased more than three hundred-fold (Parsons's largest pre-World War I machine was a 25 MW unit for Chicago) and efficiency fivefold (large pre-World War I turbines converted about twenty five percent of the steam's energy into electricity). At that time, the best steam engines had much smaller capacities and their thermal efficiency was of the order of fifteen percent: their era was clearly over.

The growth of steam turbine capacities stagnated between the two world wars, but resumed during the late 1940s. The largest machines eventually surpassed 1 GW (nearly 200,000 times larger than Parsons's first working model) and their efficiency approached, or even slightly surpassed, forty percent. Steam turbines also became the most powerful continuously working stationary prime movers for transport. Parsons himself demonstrated their advantage, when his experimental vessel *Turbinia*, 30 m long and driven by a 715 kW turbine, outran every military vessel at a grand Naval Review at Spithead, on June 26, 1897. Six years later, two score British naval ships were powered by steam turbines, soon followed by ships that became icons of the Golden Age of trans-Atlantic crossings: the *Mauretania*, the *Lusitania*, the *Olympic*, and the ill-fated *Titanic*. Subsequently, diesel engines and, more recently, gas turbines, captured the major share of the marine propulsion market but steam turbines are still common on vessels ranging from the American Nimitz-class nuclear-powered aircraft carriers to tankers that transport liquefied natural gas. Smaller stationary steam turbines also power industrial centrifugal pumps and compressors.

Coal's share in the aggregate global use of commercial energy stood at ninety-five percent in 1900, sliding below fifty percent only during the late 1950s. But as coal's relative importance declined, its absolute production grew nearly six-fold during the twentieth century, from 825 million tonnes to 4.725 billion tonnes. However, the gain was smaller in energy terms, as the average quality of extracted coal fell by more than twenty percent. The collapse of the U.S.S.R. led to a sharp drop of coal extraction in all the countries of the former Soviet Empire, and the 1990s also saw the virtual end of the venerable British coal industry. But China's extraction soared, from 1.38 billion tonnes in the year 2000 to 3.97 billion tonnes in 2013 before pulling slightly back. By 2015 the global output of bituminous (hard) coal was just above seven billion tonnes, and poorer brown coals added about 700 million tonnes. When reduced to a common energy denominator, coal provided about thirty percent of the world's primary energy needs.

In 2015, China, with 3.75 billion tonnes, was well ahead of the U.S. (over 800 million tonnes) as the world's largest hard coal producer. India (677 million tonnes), Australia (485 million tonnes) and South Africa (above 252 million tonnes) completed the group of the world's top five hard coal producers, extracting roughly seventy five percent of the global output. Germany and Russia remained the leading producers of brown coal. Only twelve other nations annually produced more than twenty-five million tonnes. The U.K. was not among them: by the year 2000 British coal extraction was only 31 million tonnes (17 million tonnes from deep mines), by 2010 deep mines produced just seven million tonnes and the last underground mine was shut down in December 2015, ending more than four centuries of an industry whose peak labor force reached 1.25 million workers in 1920 and the peak output rose to 287 million tonnes in 1913. In fact, the U.K., together with Germany, has become one of the world's leading coal importers, with the U.K. buying about 25 million tonnes and Germany importing 54 million tonnes in 2015. Coal extraction thus became restricted to fewer nations than crude oil production and most countries outside Europe and North America do not use any coal.

The modern coal industry differs in every respect from the practices that prevailed before World War I, and many were reinvented after 1950. Extraction has been transformed by the almost complete mechanization of cutting and loading, the greatly increased size of mining operations and processing designed to meet specific market needs; transport has become more economical, due to mechanical handling, and the use of special trains and larger vessels; and, a few national exceptions aside, coal now has just three principal markets, electricity generation, coke production, and cement production. The mechanization of underground mining has boosted productivity from less than one tonne/man-shift in 1900 to more than three tonnes in the best mines, reduced the labor needed, drastically cut fatalities (in the U.S. by more than ninety percent since 1930, although deadly accidents remain unacceptably common in China and the deep mines of the eastern Ukraine), and largely done away with the traditional room-and-pillar technique of coal extraction.

The room-and-pillar mining method left at least half of a coal seam behind, as it created a pattern of corridors and supporting pillars and has been replaced, wherever the thickness and layout of seams make it possible, by longwall extraction. This technique uses machines, protected by movable steel supports, to produce an advancing face of cut coal and can recover more than ninety percent of coal from the mined seam.

These trends were even more pronounced in surface coal extraction, which exploits coal seams in open mines after the removal of the relatively shallow overlying rock. Opencast mining is a safer and much more productive way of coal production, which became more common after 1970 in all but a few coal-mining countries. Surface extraction now dominates U.S. output (about sixty-five percent of the total) and produces nearly half of Russia's coal. China is the only coal superpower with a small (about ten percent) share of opencast mining. At the beginning of the twentieth century, shallow surface mines had overburden/seam ratios of just 1:2, and by its end there were many operations uncovering layers four to five times thicker than the underlying seams: giant earth-moving machines are needed for this work. Many surface mines now produce over ten million tonnes a year, with productivities averaging more than thirty tonnes/man-shift.

Highly mechanized coal preparation (processing before combustion) includes sorting, sizing (screening), and washing. Coal is often moved only a short distance to the adjacent, and appropriately named, mine-mouth power plants. Coal from fields far away from major markets is loaded on to special trains, whose large-capacity wagons (up to 100t, usually with more than a hundred linked together) remain permanently coupled to powerful locomotives and the entire assembly peddles non-stop between mine and power plant. Large coal barges are used on major rivers. Overseas shipments (Australia, Indonesia, and South Africa are leading exporters) take place in large (some more than 100,000 deadweight tonnes) bulk carriers. While coal is still commonly used by households in parts of China and India this market has virtually disappeared in both North America and Europe, as has its once very large use both in rail and water transport and the production of industrial heat.

The smallest, but most rapidly increasing of the three remaining major coal markets is cement production. About 0.1 tonne of coal are needed for a tonne of cement and the global output (4.1 billion tonnes in 2015, dominated by China's huge production) now requires more than 400 million tonnes of coal. Coke for blast furnaces is a much larger market: in 2015, its global production used just over one billion tonnes of coal, about fifteen percent of all extracted hard coal, despite the fact that iron smelting now needs more than sixty percent less coke per tonne of hot metal than it did in 1900. The demand for coke is being further reduced by more extensive steel recycling and by the direct injection of pulverized coal into blast furnaces.

The generation of electricity is the only steadily expanding market for coal. More than forty percent of the world's electricity now originates from coal-fired plants, with the share much higher in the leading coal-producing nations: more than ninety percent in South Africa, more than eighty percent in Poland, more than seventy percent in China and Australia, while the German share is now about forty percent and the U.S. share has recently declined to a third. The first small, inefficient, coal-fired stations were located in cities; after World War I they moved to the outskirts, and most large post-World War II stations were built near open-cast or major underground mines. Their most visible environmental offence, the emission of particulate matter, was eliminated, from the 1950s, by the installation of electrostatic precipitators: these simple electromechanical devices (which electrically charge the fly ash particles and collect the dust on large metal plates for disposal) capture as much as 99.9 percent of the ash. Controlling sulfur dioxide emissions proved to be a greater challenge. Only flue gas desulfurization (the wet or dry capture of the gas by reacting it with finely milled lime or limestone, forming calcium sulfate), a process commercialized during the 1970s, offered an effective way to remove most of the gas at an acceptable cost. Reducing the emission of nitrogen oxides, generated mostly by high-temperature decomposition and the oxidation of atmospheric nitrogen (coal nitrogen levels are very low), is even more costly.

By far the most intractable long-term challenge of coal combustion is the high levels of emission of carbon dioxide, the principal anthropogenic greenhouse gas. The combustion of hard coal produces about thirty percent more carbon per unit of energy than refined oil products, and nearly eighty percent more than the burning of natural gas. Coal's future has become even more uncertain thanks to increasing supplies of natural gas (now displacing coal-fired electricity generation) and there is little doubt that it will be critically dependent on the progress of global warming: rapid temperature increases would, almost certainly, accelerate coal's retreat – unless, of course, we succeed in developing new methods of advanced clean combustion or come up with acceptable ways of carbon capture and sequestration (capturing the gas and storing it underground in exhausted hydrocarbon reservoirs or in porous rocks).

During the early 1960s, crude oil became the world's most important fossil fuel, and natural gas use has been increasing faster than the consumption of petroleum: we now live in a necessarily short-lived energy era dominated by inexpensive liquid hydrocarbons, mixtures of long-chained organic molecules, which contain about eighty-five percent of carbon, the remainder being hydrogen and traces of sulfur, nitrogen, and heavy metals.

Crude oil: internal combustion engines

Like coal, crude oil has been known since antiquity. There were many locations (particularly in the Middle East) where oil seeps, pools, and tar ponds indicated its subterranean presence but its only documented use was for heating the (late) Roman baths in Asia Minor. Modern extraction of crude oil was stimulated by the search for a cheaper illuminant to replace the expensive, and increasingly scarce, oil rendered from the blubber of sperm whales, a perilous pursuit that was immortalized in Herman Melville's (1819–1891) masterpiece, *Moby Dick*. Kerosene, a colorless and highly flammable liquid that separates from crude oil between 150 and 275°C, fitted these requirements (today processed, and cleaner, kerosene is the principal aviation fuel), and helped to prevent the total extinction of one of the world's most massive mammalian species.

The modern oil industry began in Baku (Azerbaijan, at that time in Czarist Russia) in 1846 with the first exploratory well, and although the oil well completed by Colonel Edwin Drake at Oil Creek in Pennsylvania on August 27, 1859 is usually seen as the beginning of the North American crude oil industry, it actually began a year earlier in Canada, with the first manually dug oil well near Black Creek in Ontario. Drake employed, as did generations of petroleum explorers after him, the ancient Chinese method of percussion drilling (introduced during the Han dynasty) where heavy iron bits attached to derricks by long cables (in China made of bamboo, in the U.S. manila) were raised (in China by men jumping up and down on levers, in the U.S. by steam engines) and then let fall into a well to shatter the rock. In 1835, China's deepest well completed by percussion drills reached one kilometer, while Drake's well struck oil at only twenty-one meters. A few decades after Drake's discovery the search for oil was underway on all continents. Notable pre-World War I discoveries included oilfields in Texas, California, Mexico, Venezuela, Romania, Russia (at Baku, now a site of renewed exploration), and the Dutch East Indies (in Sumatra, Indonesia). The first major Middle Eastern strike was in Iran in 1908 (the Masjid-e-Suleiman field, still in production).

American cable drillers were eventually completing wells more than two kilometers deep, but the technique was doomed once improved rotary drilling rigs began to be used, after World War I. These use a heavy rotating circular table (originally driven by steam engines, later by diesel or electric motors) with drill pipes inserted in its middle. Sections of threaded drill pipes are added as the well goes deeper, but as drilling progresses the entire pipe assembly must be withdrawn to replace the drill bit. Early rotary drills used fishtail and circular-toothed drills, which were only effective in soft substrates. In 1908, Howard Robard Hughes Sr. (1869–1924) revolutionized the industry with his invention of the rotary cone drill. His company subsequently introduced many improved designs (including a smoother working tricone bit), and a century later (now as Baker Hughes) it continues to be a major maker of drilling equipment. In rotary drilling, drilling mud (it can be a water-, oil-, or synthetic-based fluid) is pumped at high pressure down the drill string and, as it flows through the bit, cools it while it removes the cuttings and exerts pressure on the well sides in order to prevent the hole from caving in. Subsequent casing and cementing stabilize the well.

CRUDE OIL TYPES AND COMPOSITION

Crude oil has a number of advantages compared to coal; the most important is its high energy density. All hydrocarbons have inherently higher energy densities than coals, because their burning oxidizes not only carbon but also hydrogen, whose combustion (producing water) releases about three times as much energy as that of carbon. Crude oils contain between eleven percent and fourteen percent hydrogen, pure methane (CH_4), twenty-five percent. Crude oils differ much less in their overall energy content than do coals, varying usually less than five percent from an average 42 GJ/t (about forty-five percent higher than high-quality bituminous coal and twice as much as good lignites).

But their density, pour point (the temperature at which the oil will flow), and sulfur content vary quite a bit and these determine their market value. The density varies because of differences in paraffin and aromatic content; the lightest (Nigerian and Algerian) crude oils have densities around 0.8 g/cm³, while many Middle Eastern (Saudi, Iranian) oils have densities as high as 0.9 g/cm³. The lightest oils, with almost no paraffin, have the lowest pour points (they flow even at below −30°C) while some waxy crude oils need heating to more than +30°C before they can even move through a pipeline. The sulfur content of crude oils goes from a trace in 'sweet' to more than three percent in 'sour' crude oils. Compared to coals, crude oils are also much easier to transport (by special trucks, barges, ocean-going tankers, or through pipelines), store (in above-ground tanks or underground cavities), and handle (they can readily be pumped on to vessels or into tanks). The capacities of tankers took off with the post-World War II worldwide expansion of crude oil shipments; typical sizes rose from less than 20,000 deadweight tonnes in 1945 to more than 300,000 tonnes by the early 1970s.

Unlike coals, crude oils are not usually burned in their natural state but are refined to yield a large variety of special fuels and non-fuel materials. Initially, refining was done by simple thermal distillation, which separates individual fractions according to their boiling points: petroleum gases boil at less than 40°C, naphthas and gasolines at 40–200°C, then kerosene (jet fuel), diesel oil, and lubricating and heavy oils at increasingly higher temperatures. The remaining solid fractions include petroleum coke, asphalt, tar, and waxes. Unfortunately, most of the world's crude oils are rather dense, and their straight thermal refining yields mostly medium and heavy fuels rather than the more valued gasoline and kerosene.

In 1913, William Burton (1865–1954) introduced thermal cracking of crude oil, in which the combination of heat and high pressure is used to break the heavier, longer-chained hydrocarbons into lighter fractions.

Refined products would have, in time, taken over many markets previously served by coal (household and commercial heating, electricity generation) but their most important use came about because of the invention of internal combustion engines. These ingenious machines created an entirely new mass market for passenger automobiles and trucks, and also replaced steam engines on railways and in water-borne transportation. Their serious development began with the large, and heavy, gas-powered two-stroke design of Nicolaus Otto, in 1866. A decade later, Otto patented a horizontal four-stroke compression engine that was still fairly heavy and relatively inefficient (about seventeen percent) and hence intended for workshops too small to have their own steam engine. However, its fundamental operating principles were the same as for all the subsequent millions of automotive and airplane engines. The first Otto-derived, but lighter and more efficient, gasoline-fueled engines were designed in 1886 and used to propel experimental carriages in Germany that year: by Karl Benz (1844–1929) in Mannheim in July (a horizontal engine on a small three-wheel chassis), and Gottlieb Daimler (1834–1900) and Wilhelm Maybach (1846–1929) in Stuttgart in the fall (a single-cylinder vertical engine mounted on a standard coach chassis). The first modern configuration for a road vehicle appeared in 1891, when the French designer, Émile Levassor (1844–1897), moved the engine from under the seats to in front of the driver: this made it possible to install larger, more powerful engines and led to the design of the protective (and soon also aerodynamic) hood. Many technical innovations of the 1890s combined to produce more powerful, more efficient and also safer engines. The replacement of dangerous open-flame hot-tube ignition, first by low-voltage magnetos and, after 1901, by Robert Bosch's new type of spark plug, was a major advance. But cars, produced by artisanal methods in small workshops, remained expensive and often unreliable and their total numbers were low; in the U.S. just 8,000 in 1900 and 78,000 five years later.

Everything changed with Henry Ford's (1863–1947) great innovations. On October 1, 1908, Ford introduced his Model T, deliberately built and priced to create a new mass market. It combined good-quality materials (heat-treated vanadium steel), practical design (including the first water-cooled four-cylinder engine and generous clearance, to negotiate not only unpaved and muddy roads but even a newly plowed field), and a price within the reach of anybody on an average salary. The car's price was further reduced with the introduction of Ford's famous assembly line in 1913. This new manufacturing arrangement relied on the subdivision of labor into a large number of simple repetitive tasks; it was subsequently as widely condemned and criticized (as numbing and dehumanizing) as it was admired, copied, and refined, not only by other carmakers but also in all kinds of other industries. The huge productivity gains of assembly manufacturing were undeniable and by 1914 Ford could reward his workers with doubled wages and shorter working hours. After World War I, the Model T, introduced at $850 in 1908, sold at $265 and by 1927, when its production ended, mass automobilization of America, and later of the world, was unstoppable.

As this new mode of passenger transport was taking shape, a different internal combustion engine was introduced by Rudolf Diesel (1858–1913), whose invention was patented in 1892. Diesel's engines work without spark plugs, as fuel injected into the cylinder ignites spontaneously due to the high temperatures generated by high (14–24) compression ratios. Early diesels were much heavier and slower than Otto-cycle machines but were inherently more efficient, converting more than forty percent of their fuel into kinetic energy. Moreover, they burn heavier, cheaper, and more energy-dense fuel. They were first used in marine propulsion; then after 1930 they appeared more frequently in locomotives, trucks, and cars. By the 1990s, most of the world's trucks were diesel-powered, and diesels now account for about forty percent of all new passenger car registrations in Europe. However, all the fastest trains are now powered by electricity.

Gasoline engines would not have been so successful without the elimination of a problem that limits the performance of Otto's combustion cycle: spontaneous ignition of the fuel-air mixture produces a pressure wave that moves in the direction opposite to that of the spreading flame and causes characteristic and destructive "knocking". In early engines, the only way to prevent this was to run them at low compression ratios (below 4.3), making them relatively inefficient. An inexpensive (but eventually highly regrettable) solution came in 1921, when Thomas Midgley (1889–1944) identified tetraethyl lead (added as a mere one-thousandth by volume) as a highly effective anti-knocking additive. Compression ratios eventually rose to 8–10.5; the additive also made it possible to run faster and more powerful aeroengines. Half a century after its introduction, leaded gasoline began to be withdrawn from the market because of its health effects (lead poisoning, particularly crippling in small children) and also because the lead would poison the platinum catalyst used in the catalytic converters that became mandatory to cut automotive emissions of carbon monoxide, nitrogen oxides, and unburned hydrocarbons.

Internal combustion engines also made possible the long-deferred dream of heavier-than-air flying machines. In 1903, after several years of theoretical studies, experiments, and glider flights, Orville (1871–1948) and Wilbur (1867–1912) Wright, both experienced bicycle mechanics, not professional engineers or scientists, built their own four-cylinder engine to power their experimental *Flyer*, which accomplished the first short successful piloted flight on December 17 on a beach in North Carolina. Aeroengines and airplanes developed far more rapidly than cars: the English Channel was crossed in 1909 and by 1914 plans were in place to cross the Atlantic via Iceland. War interrupted these designs but fostered the development of engines and structures for new fighter and bomber planes. Commercial aviation began in the early 1920s, the Atlantic was crossed for the first time in 1919, and just two decades later, PanAm's luxurious Clipper (a Boeing 314, with dining room and beds) began staged trans-Pacific crossings from San Francisco to Hong Kong.

Oil and natural gas: hydrocarbons dominant

Before World War II the U.S. was the only industrialized country with a significant share of hydrocarbons in its primary energy supply (oil and gas provided forty-five percent of the total in 1940). Coal remained the dominant fuel in post-World War II Europe throughout the 1950s but the continent's subsequent transition to oil- and gas-based economy was rapid. As Western Europe had no major sources of hydrocarbons it became a major importer of crude oil from North Africa and the Middle East. Japanese oil resources were even more minuscule, and the country's high rates of economic growth made it the world's largest petroleum buyer during the 1960s. The discovery of the giant Groningen natural gas field, in the Netherlands, the development of the North Sea oil and gas fields (the first productive wells were drilled during the 1960s), and imports of Siberian gas to Central and Western Europe (begun in the early 1980s) accelerated the continent's transition to hydrocarbons, leading to either the complete demise (in the Netherlands), or the drastic reduction (in France, the U.K., and Germany) of coal mining. Most of the newly-industrializing economies of Asia and Latin America bypassed the coal stage entirely (even those with sizable domestic coal deposits) and based their development on local or imported oil.

This rising demand could not have been met without widespread and sometimes truly amazing technical innovations. The post-World War II production of hydrocarbons benefited from new techniques, ranging from three-dimensional computerized geophysical exploration, to better refining methods, and new modes of gas transportation. Recovery of oil, traditionally less than a third of all liquids present, improved with the introduction of horizontal drilling during the 1980s. Horizontal wells can intersect and drain multiple fractures within, and penetrate larger volumes of, the oil-bearing rock: many can produce two to five times as much as vertical or slightly deviated wells drilled into the same reservoir. By the 1990s, the longest horizontal wells reached four kilometers, while vertical wells are now frequently drilled to depths below five kilometers. And during the first decade of the twenty-first century horizontal drilling was combined with hydraulic fracturing to start a new era of American oil and gas renaissance. The technique, commonly known as fracking, forces a mixture of pressurized water, sand and various chemicals into wells drilled horizontally through shale deposits and liberates oil or natural gas present in the rock.

Another key post-World War II innovation was the development of an entirely new set of production techniques for offshore drilling. The first well out of sight of land was completed off Louisiana in 1947; seventy years later, offshore wells supply nearly a third of the global oil output. Assemblies used for offshore drilling include fixed (jack-up) rigs, submersible and semi-submersible platforms, and drilling ships (some working in waters more than two kilometers deep). Offshore production platforms built above the largest undersea fields are among the most imposing structures anywhere: some weigh as much as nuclear aircraft carriers and rise more than a hundred meters above sea level.

Major changes in refining began in 1936 when Eugène Houdry introduced a new catalytic cracking process and its subsequent improvements were ready just in time to produce the large quantities of gasoline needed for vehicles and airplanes during World War II. The greatest post-World War II improvements in oil refining were the use of a synthetic zeolite (a crystalline aluminosilicate) catalyst that boosted gasoline yields by up to fifteen percent, and the introduction of hydrocracking, a technique that combines catalysis with hydrogenation and gives higher gasoline yields. The demand for refined products is nation-specific: the U.S. and Canada need high shares of gasoline (almost half the total), while in Japan half the final demand is middle distillates and fuel oil for industrial and household heating. Non-fuel fractions (lubricants and paving materials) account for about twelve percent of the initial mass of refined crude oils.

Worldwide exploration for hydrocarbons has dramatically changed the distribution of oil and gas reserves. The first Middle Eastern super-giant oilfield, Kuwaiti al-Burgān (still the world's second largest repository of crude oil), was discovered in 1938. Saudi al-Ghawār, the world's largest oilfield (containing almost seven percent of the world's oil reserves), was drilled in 1948. At the time the Organization of Petroleum Exporting Countries

(OPEC, set up in 1961 by Saudi Arabia, Iran, Iraq, Kuwait, and Venezuela and later expanded to thirteen members) decided to quintuple its crude oil price (between October 1973 and April 1974), the Middle Eastern oilfields (principally in the Persian Gulf countries of Saudi Arabia, Iran, Iraq, Kuwait, and Dubai) contained some seventy percent of all oil reserves. OPEC's extortionate pricing was finally broken after a second round of steep price rises during 1979 and 1980, triggered by the fall of the Pahlavi dynasty in Persia and the takeover of the government by fundamentalist mullahs. Temporary fluctuations aside, world oil prices remained generally low and stable from 1985 to 2003.

A subsequent rise brought them to a new record level in 2008 followed by a large drop induced by the worst post-World War II economic crisis; after a quick recovery the prices reached a new high by 2013 followed by another steep drop. When adjusted for inflation, prices during 2016 were below the level of the late 1990s. Further fluctuations must be expected. Although there has been no substantial shift in the distribution of conventional (liquid) oil reserves (they remain extraordinarily concentrated in the Persian Gulf region), the U.S. has become, once again, the world's largest crude oil producer thanks to the large-scale deployment of hydraulic cracking. These new supplies, combined with reduced demand in slowly growing Western economies as well as in China, have undermined OPEC's ability to set the world oil price.

Despite an enormous expansion of oil extraction (from about twenty million tonnes in 1900 to 500 million tonnes in 1950, 3.2 billion tonnes in 2000 and 4.3 billion tonnes in 2015), the global reserve/production ratio of crude oil has been higher during the past decade than it was fifty years ago. Oil extraction is now more widespread than coal mining: more than forty countries are substantial producers, of which the top five (in 2015 they were U.S.A, Saudi Arabia, Russia, Canada and China) extract nearly half (forty eight percent) of the global total. In 2015 crude oil supplied a third of the world's commercial energy and fueled more than ninety percent of the world's transport (it supplies all energy for flying). Transport needs would have been much higher without the many technical advances that resulted in lighter, more reliable, and more energy-efficient internal combustion engines – and without the invention of an entirely new prime mover, the gas turbine.

GAS TURBINES

These remarkable machines, invented independently during the 1930s by Frank Whittle (1907–1996) in England and by Hans Pabst von Ohain (1911–1998) in Germany, were put first into service in military jets just before the end of World War II. Their rapid development saw them powering airplanes that surpassed the speed of sound (first on October 14, 1947) and soon afterwards began to transport passengers at slightly sub-sonic cruising speeds (about 900 km/h). The British Comet, a pioneering jetliner, introduced in 1952, had to be withdrawn after several fatal accidents (caused by faulty fuselage design) and by the time it was reintroduced it had been overtaken by the Soviet Tupolev 104 (in 1956) and the American Boeing 707 (in 1958). The airplane that revolutionized intercontinental travel, the wide-bodied Boeing 747, flew for the first time in 1969; hundreds of jumbos (mostly the advanced 747–400 version) are still in service today. Boeing's most successful airplane has been the baby of its line-up: by 2016 the company had built about 9,300 737s (first flown in 1967), making it the most popular jetliner in history. In contrast, the extravagant British-French supersonic Concorde (launched in 1969, withdrawn in 2003) was always an uneconomical, super-expensive, and very noisy oddity, limited to a few routes. No single figure demonstrates the post-World War II expansion of commercial flying as well as the global annual aggregate of passenger-kilometers (p-km): in 1950 this was forty billion, by 2000 it passed three trillion, and by 2015 it reached about five and half trillion, nearly a one-hundred-forty-fold increase in sixty-five years.

Gas turbines have also found a variety of stationary applications: they are the preferred choice to power the large centrifugal compressors that push natural gas through pipelines and supply the pressure needed for many chemical and metallurgical

GAS TURBINES (*cont.*)

processes, and they are increasingly used to generate electricity in relatively small, decentralized, facilities. Technical improvements have lifted the efficiencies of these machines above forty percent and in combined cycles (using the exiting hot gas to heat water for a smaller steam turbine) they were the first converters whose efficiency surpassed sixty percent.

Arguments about an imminent peak of global oil production and the coming end of the oil era have become common at the beginning of the twenty-first century, but they have been unjustifiably pessimistic. While the chances of discovering new supergiant oil fields have certainly become very low, there remains an enormous potential for individually smaller, but cumulatively substantial, new discoveries in parts of the Middle East, Siberia, Central Asia, Africa, and in many offshore basins. In addition, there are huge deposits of non-traditional oil, some (Canadian tar sands in Alberta, Venezuelan heavy oil) already exploited commercially, and American hydraulic fracturing opened the commercial way to recover significant shares of crude oil locked in abundant shales. Consequently, global civilization will rely on crude oil for decades to come, while continuing to use more natural gas, the simplest, cleanest, and hence in many ways most desirable, of all hydrocarbons.

The only well-documented use of natural gas in a pre- industrial society was its burning to evaporate brine in landlocked Sichuan, in China, which began in the early Han dynasty (about 200 B.C.E.). Natural gas is composed mostly of methane, with small amounts of ethane, propane, hydrogen sulfide, and nitrogen. It is often mixed with crude oil in the same reservoir. During the early decades of the oil industry (when there were no high-pressure long-distance pipelines) this so-called associated gas had to be burned off, if there was no way to use it locally. This wasteful practice is still common in some hydrocarbon-producing regions of Africa and the Middle East; the sites are marked by enormous bright spots on night-time satellite images (as bright as the light produced by some large cities).

Throughout the Western world, this waste of a precious resource was almost completely eliminated with the development of long-distance natural gas pipelines, and gas became the most sought-after choice for space heating and many industrial processes. Pumping compressed natural gas through a pipeline takes more energy than moving the equivalent mass of crude oil, but its high quality justifies long pipelines and even some expensive (but relatively short) undersea links. The U.S. was the first country with extensive gas pipeline networks, mostly originating in Texas, Oklahoma, and the Gulf of Mexico. After World War II Canada became a major gas exporter to the U.S., and new pipelines crossed also into Mexico, creating a continent-wide system.

Natural gas now provides nearly a quarter of the world's primary commercial energy. The best available accounts show that its reserves contain about two-thirds of energy available in crude oil reserves, roughly 155 as against 240 billion tonnes of oil equivalent in 2015. Despite the tripling of natural gas extraction from 1975 to 2015, the worldwide reserve/production ratio of the fuel is nearly sixty years, compared to just over fifty for crude oil. Iran and Russia have the largest reserves, followed by Qatar, Turkmenistan and the U.S.. The quest for cleaner electricity generation and more affordable availability of intercontinental shipment of liquefied natural gas will increase the fuel's future share in global primary energy supply.

Electricity: the first choice

You do not have to have a discriminating knowledge of different energy sources to know that this section's title rings true. Pushing or flipping a switch is all that is required to start the flow of this most convenient, most flexible, and most useful of energies.

There is no need to store it, no pouring of liquids into tanks, stacking or shoveling of solids into basements or sheds, kindling, stoking or tending, removal of ashes, or cleaning of pipes. And yet this marvelous form of energy makes it possible to heat, cool, or light interior spaces, power countless motors that perform tasks ranging from keeping premature babies alive in their incubators to circulating blood in machines during heart by-pass surgery or, to choose an entirely different and randomly mismatched duo of examples, from centrifugally separating cream from milk to propelling trains at more than 300 km/h.

The list of positives goes on: electricity is noiseless and aseptically clean at the point of use (I will describe the polluting effects of large power plants shortly), its availability is instantaneous, its cost has been reduced by technical advances to a small fraction of typical income, and its flow can be controlled with unmatched precision to supply the speeds or heat levels needed by a myriad industrial, transportation, and household processes. Electricity can be converted not only to heat and motion (the former conversion is one hundred, the latter, with large motors, more than ninety percent efficient) but also to light and chemical potential energy: the only major sector of the modern economy where it is absent is air transport. Although in 2016 an experimental plane, whose propellers were powered by photovoltaic electricity, circumnavigated the world, it carried just two pilots – while modern jetliners carry hundreds of people as well as cargo. And because it can generate temperatures higher than the combustion of any fuel, electricity is also ideal for many metallurgical and other high-temperature processes.

There is no better proof of electricity's importance than a couple of simple thought experiments: list your everyday tasks and actions that depend on electricity, or write down the objects, tools, machines, services, and processes that would be absent in modern society without its presence. Even confined just to housing, the latter exercise reveals poorly illuminated rooms (lit by dripping candles, smelly kerosene, or coal gas), spoiled food (no easy refrigeration), tiring walks to the nth floor (no elevators), the laborious washing and ironing of clothes, and, of course, no telephones, nor anything electronic: no radios, stereos, televisions, or music players, and no surfing of the web. On a dairy farm you would face endless manual chores, such as pitchforking hay, chopping and grinding the feed, pumping water and pouring it into troughs, and milking the animals. In pre-electric steam-driven factories you would work under ceilings crammed with iron or steel line shafts, connected by belts to parallel countershafts which powered individual machines. A mishap anywhere along this complicated path of power distribution (leaking boiler, damaged shaft, slipped belt) would shut down the entire assembly – and, an opposite inconvenience, the entire assembly would be running even when only a few machines were needed. Accurate speed adjustment would be impossible with so many belt transmissions and the whole set-up noisy, dangerous, and dim. Fortunately the replacement of shafts and belts by electric motors powering individual machines brought precise control, obviated the need to operate entire sets, and uncluttered ceilings either to let in natural light or carry ample electric lighting.

Combustion of fossil fuels, and the use of high-temperature (pressurized) steam to drive turbogenerators has been the globally dominant means of electricity generation since the industry's commercial beginnings, when in 1882 Thomas A. Edison completed the world's first two small coal-fired stations in London (at Holborn Viaduct) and New York (on Pearl Street near the city's financial district). The rapid growth of typical power plant capacities was made possible by a combination of several key inventions: steam turbines, transformers, the conversion of direct to alternating current, and high-voltage transmission, and by a process of continuous innovation and efficiency improvements. Combustion of fossil fuels now produces about sixty-three percent of the world's electricity, and the best efficiencies of the entire process are about forty percent.

THERMAL POWER PLANTS

No matter what their fuel, thermal power plants (a common, though inaccurate, term for electricity-generating stations; one that also includes nuclear power stations, where fission, rather than combustion, provides the requisite heat) share a boiler and turbogenerator arrangement. A boiler is a large chamber, whose walls are lined with steel tubes, fed with demineralized and pressurized water, and heated by combustion of the fuel injected into the chamber, mixed with preheated air. About ten percent of the heat released by combustion escapes through a tall chimney and carries off incombustible particulate matter (ash) and combustion gases, mainly water vapor, carbon dioxide, and sulfur and nitrogen oxides (Figure 21). The steam produced in the boiler is at a temperature of more than 550°C, and it is led to a turbine, where its expansion pushes the blades and rotates a generator in a magnetic field to produce alternating current. The capacities of fossil-fueled turbogenerators rose from 1 MW in 1900 to nearly 1.5 GW before their growth stopped during the 1970s.

The steam that leaves the final (low-pressure) stage of a turbine is condensed in a vacuum within a condenser, and this accounts for most of the heat lost during electricity generation (nearly five times as much as from the chimney). The condensed water is, after preheating, returned to the boiler. The cooling water that is heated in condensing the steam is released into a stream or other water body and new cold water pumped in. But because this release raises the downstream water's temperature (endangering some aquatic species, but benefiting others) even power plants located in regions with a plentiful water supply now cool the condenser water for reuse. The giant concrete cooling towers are, besides the tall chimneys, usually the first objects indicating the distant presence of a thermal power plant.

Coal-fired stations must also have large on-site fuel storage, a mill to pulverize it to less than 0.3 mm so it can be blown into a boiler, and bulky fly-ash precipitators, attached to the chimneys, which remove all but a tiny amount of the smallest particulates.

THERMAL POWER PLANTS (*cont.*)

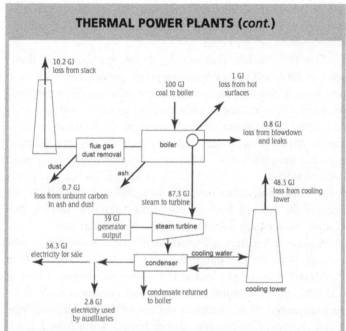

Figure 21 The energy balance of a coal-fired electricity generating plant

Many coal-fired stations also have desulfurization units, associated facilities for preparing limestone, and nearby ponds for storing sulfate slurry. These facilities commonly use from two to four percent of the plant's electricity generation, resulting in a final thermal efficiency of no more than forty percent.

Electricity: beyond fossil fuels

About a third of the world's electricity is not generated by the combustion of fossil fuels. This contribution comes largely from two very different processes, the conversion of falling water's kinetic energy, and the generation of heat through the fission of an isotope of the heaviest stable element, but recent decades have also seen rapid increases of electricity generated by wind turbines and photovoltaic (PV) cells.

The first, widely distributed process, encompassing thousands of dams and hydrogenerating plants, provided about sixteen percent of global electricity in 2015, the other has been restricted to about 440 operating nuclear reactors in thirty one countries and added up to about eleven percent of all electricity generation in 2015. Falling water began to produce electricity at almost the same time (the early 1880s) as steam-power. State-supported construction of large hydro stations took off during the 1930s, with major projects in both the U.S. and the U.S.S.R. (there inspired by Lenin's famous, but badly mistaken, dictum that Communism equals Soviet power plus electrification).

American projects included such iconic structures as the Hoover Dam on the Colorado, near Las Vegas (completed in 1936), and the Grand Coulee, the continent's largest station at 6.18GW, on the Columbia River, in full operation since 1941. As Europe and North America ran out of suitable sites for large hydro projects, dam building shifted to Asia, Latin America, and Africa. At the beginning of the twenty-first century, the only countries that did not generate any hydroelectricity were in the most arid parts of the subtropics. In 2015 nearly forty countries derived more than fifty percent of all their electricity from this renewable resource, and the five largest producers, China, Canada, Brazil, U.S. and Russia produced nearly sixty percent of the global total.

Record hydroelectric dam-building achievements include height, 335m (the Rogun dam on the Vakhsh in Tajikistan), reservoir area, 8730km^2 (behind the Akosombo dam on the Volta, in Ghana, an area nearly as large as Lebanon), and capacity, 22.5GW (Sanxia, the Three Gorges dam across the Yangtze in Central China). This huge project epitomizes the problems that have led to increasing opposition to large hydroprojects:

flooding of settlements and farmland requires relocation of large numbers of people; Sanxia displaced nearly 1.5 million, and worldwide estimates put the total of displaced people at no less than forty million, and perhaps twice as many, during the twentieth century.

The silting of reservoirs, particularly common in monsoonal Asia, and especially rapid in areas where heavy deforestation has removed much of the protective forest cover in a dam's watershed, shortens the useful economic life of these expensive storages. Large reservoirs also cause considerable aging of the average river runoff (in some of the world's largest rivers water entering the sea can be from six months to more than one year old, compared to the average of two to four weeks for streams without any impoundments), and lower downstream temperatures. Tropical reservoirs are also often invaded by aquatic weeds and can be a major source of greenhouse gases (carbon dioxide and methane from decaying vegetation), which weakens the promotion of hydroelectricity as an environmentally benign form of generation.

Nuclear generation had its genesis in the advances of pre-World War II physics and the wartime quest for fission bombs; this effort was guided by the fear that Nazi Germany would develop its own nuclear weapons. The possibility of nuclear fission was demonstrated for the first time in Germany, in December 1938, by Otto Hahn (1879–1968), Lise Meitner (1878–1968) and Fritz Strassmann (1902–1980), after they irradiated uranium with slow neutrons and found isotopes other than the transuranic elements which had formed in previous experiments. The first sustained chain reaction took place at the University of Chicago on December 2, 1942, and the first fission bomb was tested at Alamogordo, New Mexico, on July 16, 1945. The two bombs that destroyed Hiroshima (a uranium bomb equivalent to 12,500 tonnes of TNT) and Nagasaki (a plutonium bomb equivalent to 22,000 tonnes of TNT) were dropped on August 6 and 9, 1945 and their immediate casualties were, respectively, around 119,000 and 70,000.

The road toward nuclear electricity generation began after World War II, when Hyman Rickover (1900–1986) began his determined push for the construction of nuclear-powered submarines. The *Nautilus*, the first of what was to become a key component of the U.S. strategic triad (long-distance bombers and land-based missiles being the other two), was launched in January 1955. Two years later, a nearly identical pressurized water reactor (PWR) went into operation in the country's first nuclear-powered electricity-generating station in Shippingport, Pennsylvania. PWRs produce steam indirectly, using two loops to minimize any accidental release of radioactivity. The entire reactor core is submerged in water inside a heavy pressure vessel; in the first loop pressurized water flows through the reactor's core, removes heat from the fuel rods (corrosion-resistant, zirconium steel-clad tubes, filled with pellets of enriched uranium dioxide), and brings it into a steam generator (basically a heat exchanger and the equivalent of a boiler), the second moves the generated steam to a turbogenerator and the condensed water back into the steam generator. The use of high pressure (in excess of thirteen megapascals) makes for a compact reactor design and both the reactor and steam generator are enclosed within a strong containment structure.

In contrast, the world's first commercial nuclear power station, Calder Hall, commissioned in October 1956, inaugurated a British reactor series cooled by pressurized carbon dioxide and whose fuel rods were clad by a magnesium alloy (hence called Magnox reactors). Yet another reactor design was adopted in Canada: in this the fuel is natural uranium and the coolant heavy water (deuterium oxide). After a slow start, scores of new stations were ordered in more than twenty countries between 1965 and 1975, and there was a widespread expectation that fission would dominate the world's electricity supply by the century's end.

These expectations were further strengthened after OPEC's first round of oil price increases in 1973–74, but cost over-runs, major delays in construction, safety concerns (made more acute by an accident at the Three Mile Island plant in Pennsylvania in 1979), absence of provisions for permanent storage of radio-active wastes (all countries store them in temporary facilities), and a sharp, post-1975 decline of new demand for electricity combined first to slow and then practically shut down any further progress of nuclear energy in the Western world. France was the only exception and its ambitious program (in 2015 it had fifty-eight reactors based on a Westinghouse PWR design that was replicated in several standard sizes) now delivers about seventy-seven percent of that country's electricity.

However, no Western reactor (surrounded by a contain-ment vessel and subject to much tighter operating proce-dures) could have released so much radiation as did the unshielded and carelessly operated Chernobyl reactor in the Soviet Ukraine. The meltdown of its core in May 1986, and the subsequent environmental and health impacts (particularly over large areas of Ukraine and Belarus) made it almost impos-sible to sell fission as a major component of future energy supply in risk-conscious Western countries. These concerns were further elevated by the disastrous failure of three nuclear reactors in Japan's Fukushima Dai-ichi station damaged by the tsunami on March 11, 2011. This led to the closure of all Japa-nese reactors and to Germany's decision to phase out nuclear generation by 2022.

But nuclear capacities continue to expand in China, India, South Korea and India and operating licenses have been extended for many older plants in the EU and North America. Besides France, the other countries deriving high shares of their electric-ity from nuclear fission are Hungary and Slovakia (more than fifty percent), Ukraine and Belgium (nearly fifty percent), and Sweden and Switzerland (about forty percent).

Geothermal electricity generation remains marginal, but PV and wind generation have been increasing very rapidly: between

2000 and 2015 the former expanded about two hundred and thirty fold, the latter about twenty-seven fold. By 2015 their combined share was still only about 4.5 percent of the world's electricity but it will continue expanding. In contrast, experimentally tested ways of electricity generation, such as wave-driven power plants, and such anticipated techniques as space-based (even Moon-based) photovoltaics are highly unlikely to make any significant contributions during coming decades.

GEOTHERMAL, WIND, PHOTOVOLTAICS

Geothermal fields have been tapped in a number of countries. Italy's Larderello was first in 1902, New Zealand's Wairakei has been operating since 1958, and California's Geysers since 1960. There are also geothermal stations in Mexico, Indonesia, and the Philippines. The U.S. has the highest installed capacity, followed by the Philippines and Italy but the global total does not even exceed 10 GW.

While geothermal capacity has been increasing very slowly, wind generation has experienced an exponential take-off since the mid-1990s: its global aggregate rose from 1 GW in 1999 to 432 GW in 2015 (Figure 22). European countries, particularly Denmark, Germany, and Spain, have led this rapid expansion, but by 2015 the highest installed capacities were in China and the U.S., followed by Germany, India and Spain. The expansion of wind-driven generation was helped by the guaranteed fixed price for wind-generated electricity and by improved turbine designs (with blades optimized for low speeds) and much larger turbine sizes: the largest capacities grew from just 50 kW in the early 1980s to more than 1 MW by 2000 and to 8 MW (for offshore turbines) in 2015. Many countries have very ambitious plans for large capacity increases by 2030.

PV capacities have been expanding at an even faster rate and by the end of 2015 reached 256 GW. China, Germany, Japan, U.S. and Italy lead the national ranking. As is the case with wind turbines, many countries anticipate further rapid growth of installed PV capacities. Central solar power stations use concentrated radiation to heat water or molten salt and can generate electricity even after sunset.

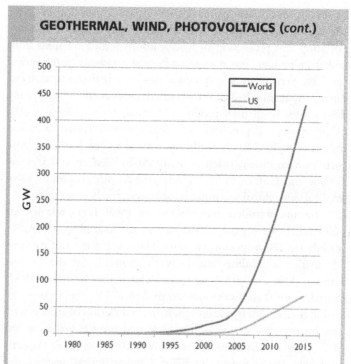

GEOTHERMAL, WIND, PHOTOVOLTAICS (*cont.*)

Figure 22 World and U.S. wind–generating capacity (plotted from statistics published in U.S. Energy Information Administration and by the World Wind Association)

Energy and the environment: worrying consequences

The extraction, transportation, processing, and combustion of fossil fuels, and the generation and transmission of electricity produce an enormous array of environmental impacts, from local to global, and from fairly short-lived degradations to long-lasting adverse changes. Local environmental degradation, caused by underground and opencast coal mining (disruption of water tables, ground subsidence, and acidic mine drainage), and by transportation and processing of hydrocarbons (crude oil spills from tankers, pipelines, and storage tanks, and accidents in refineries) can have very serious, and sometimes long-lasting, impacts, but during the second half of the twentieth century the world's rising demand for fossil fuels produced a much more worrying class of environmental problems, affecting not only entire cities and industrial regions (such as photochemical smog), but even large parts of a continent (such as acid rain in Western and Central Europe and Eastern North America) and with profound global implications (global warming is, of course, the best example).

The uncontrolled combustion of fossil fuels releases air pollutants, from tiny particles of carbon and incombustible matter (mostly oxides) to poisonous gases. The combination of particulate matter and sulfur dioxide was responsible for the famous London (U.K.) smog whose worst episode (in December 1952) caused more than 4,000 premature deaths. The replacement of coal in households by cleaner fuels, the installation of electrostatic precipitators, use of low-sulfur fuels, and the gradual adoption of flue gas desulfurization did away with smog in all Western countries (but it persists in many Chinese cities). Instead, the world's major urban areas now experience recurrent episodes of

MAJOR AIR POLLUTANTS

Particulate matter (PM) includes all solid or liquid aerosols with a diameter of less than 500µm. Large, visible particulates, including fly ash, metals, dust, and soot, originating from uncontrolled combustion in household stoves and industrial and power plant boilers, settle fairly rapidly and are not usually inhaled. In contrast, aerosols with diameters below 10µm can be inhaled, and those with diameter 2.5µm and smaller travel deep into the lungs and are responsible for a high incidence of chronic respiratory diseases.

MAJOR AIR POLLUTANTS (*cont.*)

PM can stay airborne for long periods of time and even be transported across oceans. Only a week after the Iraqis set fire to the Kuwaiti oil wells in February 1991, PM from these fires arrived in Hawaii.

Sulfur dioxide (SO_2) is a colorless and, at low concentrations, odorless gas; higher concentrations produce a characteristic pungent and irritating smell. Its origin is in the oxidation of sulfur, present in coals and crude oils (usually between one and two percent of their mass) and, to a much lesser extent, in gasoline and natural gas. Nitrogen oxides (NO and NO_2 commonly labelled NOx) are produced when combustion at high temperatures (above all in coal-fired electricity-generating plants and internal combustion engines) splits the molecules of atmospheric nitrogen (N_2), leaving the nitrogen atoms free to combine with oxygen. Hydrocarbons are emitted during incomplete combustion of fuels and volatile fractions can also escape from fuels, particularly during the processing, distribution, and sale of petroleum products.

Incomplete combustion also produces carbon monoxide (CO), a colorless and odorless gas, emitted by all kinds of internal combustion engines (in cars as well as in lawn mowers or chain saws) and by smoldering coal seams or garbage fires. In the presence of sunlight the combination of NO_x, hydrocarbons, and CO feeds a series of reactions, which produce photochemical smog, whose first sign is a significant reduction in visibility. Eventually, the reactions produce high levels of ozone and other highly reactive gases, responsible for a variety of health effects (eye irritation, allergies, and aggravation of respiratory illnesses), and damage to crops, trees, and some materials (rubber, plastics, textiles).

photochemical smog, resulting from chains of atmospheric reactions of nitrogen oxides, hydrocarbons, and carbon monoxide.

Oxides of sulfur and nitrogen emitted from the combustion of fossil fuels are gradually oxidized to produce atmospheric sulfates and nitrates, which lower the acidity of precipitation to well below the normal level (pH 5.6, caused by the permanent presence of carbon dioxide) and create acid rain and fog. As long as pollutants were emitted close to the ground, from houses and low chimneys, acid rain remained a local affair. As coal-fired electricity generation, mostly in large plants with very tall chimneys, increased after the 1950s, pollutants began to be carried up to a thousand kilometers downwind, and acid rain began to affect areas far from the source of the emissions. Parts of Europe, including southern Scandinavia, the Netherlands, Germany, Poland, and the Czech Republic, experienced particularly high levels of acid rain from the late 1960s, and it affected the eastern third of the U.S. and neighboring parts of Canada from the late 1970s. Acid rain damages sensitive fish and amphibians, leaches alkaline elements, releases aluminum and heavy metals from soils, and has both acute and chronic effects on the growth of coniferous forests; it also corrodes exposed steel structures, and damages stone (limestone above all), paints, and plastics.

Fortunately, early concerns about the progress of acid rain-induced environmental impacts were not realized, thanks to the combination of cleaner fuels (more natural gas and low-sulfur coal), and the commercial desulfurization of many large coal-fired electricity-generating plants. As a result, during the last two decades of the twentieth century, the emissions of sulfur oxides declined in Europe, the U.S., Canada, Japan, and (due to the installation of flue gas desulfurization in large coal-fired power plants) since 2008 also in China, now the world's largest consumer of coal. Another major control success has been the drastic reduction of the three precursors of photochemical smog, emitted by car engines, thanks to three-way catalytic converters, first introduced in the U.S. during the early 1970s. Compared to pre-control levels, these devices cut specific emission (g/km) by more than ninety-five percent for hydrocarbons and CO, and by about ninety percent for NO_x, and further improvements lie ahead.

There are many other environmental problems directly or indirectly connected to the world's rising energy demands.

Some, like high rates of severe respiratory problems among rural populations in Asia and Africa that rely on unvented (or insufficiently vented) indoor combustion of wood and crop wastes, have ready technical solutions, through the introduction of better stoves and the eventual adoption of modern fuels. Others, like excessive silting of reservoirs that impound water for electricity generation, have no easy fixes. Some, such as objections to the visual impact of new large-scale wind-turbine installations, may seem frivolous; others, such as the continuing inability of any country with a substantial nuclear generation capacity to come up with a safe solution to the long-term (thousands of years) disposal of radioactive waste, reflect not so much our lack of technical ingenuity (we know how to store these wastes) but rather the increasing inability of the public decision-making process to come up with timely and effective policies.

But none of these challenges is as difficult, and the failure to cope with it as potentially costly, as the continuing increase of anthropogenic greenhouse gases that could lead to a rapid global warming. This intractable problem is not solely due to energy consumption, but combustion of fossil fuels is the largest source of CO_2, by far the most important anthropogenic greenhouse gas. Until about 1910, the largest source of CO_2 attributable to human actions was changing land use, above all the conversion of forests and grasslands into crop fields, and the destruction of vegetation to build new cities, industrial plants, transport links, water reservoirs, and open new mines. By the beginning of the twenty-first century these changes (led by tropical deforestation) continued to release annually about four billion tonnes of CO_2, but nine times as much was emitted by the combustion of fossil fuels. As a result, by 2015 the average atmospheric levels of CO_2 (just over 400 parts per million or 0.04 percent) were more than forty percent higher than during the pre-industrial era – and may be another forty percent higher by the century's end.

CO_2 CONCENTRATIONS AND OTHER GREENHOUSE GASES

An ingenious technique has made it possible to very accurately reconstruct CO_2 atmospheric levels going back for nearly one million years: the analysis of tiny air bubbles preserved in the ice of Antarctic and Greenland glaciers (removed by drilling deep cores) shows that, during the five thousand years of pre-industrial civilization, atmospheric CO_2 concentrations fluctuated within the narrow range of 250–290 parts per million (ppm, the same as 0.025–0.029 percent). They rose from about 280 ppm in 1850 to 320 ppm in 1958, the first year of systematic measurements at two American observatories, Mauna Loa in Hawaii and the South Pole (Figure 23). In November 2015 the Mauna Loa readings, taken as the standard indicator of the planet's steadily rising background CO_2 level, surpassed 400 ppm for the first time.

Although CO_2 remains the most important anthropogenic greenhouse gas, its relative contribution to possible global

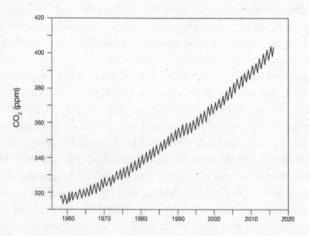

Figure 23 Atmospheric CO_2 concentrations (plotted from data available from the Carbon Dioxide Information and Analysis Center)

CO₂ CONCENTRATIONS AND OTHER GREENHOUSE GASES (*cont.*)

warming has steadily declined, and those of other greenhouse gases generated by human activities increased. Methane (CH_4) leaks from natural gas wells, pipelines, and coal mines, is emitted from landfills and rice paddies, and is produced, in relatively large volumes, by the digestion of feed in ruminant animals. Nitrous oxide (N_2O) comes from bacteria-mediated transformations of nitrogen fertilizers, from industries, and from fossil fuel combustion. Ozone is a major constituent of photochemical smog. And before they were banned, by international agreement, during the late 1980s, chlorofluorocarbons (CFCs) escaped from refrigerators, air conditioners, and industrial processes. All these gases have atmospheric concentrations much lower than CO_2 but their relative ability to absorb the outgoing infra-red radiation is much higher. Consequently, they now account for thirty–five percent of the total anthropogenic radiative forcing (warming) that was just short of 3 W/m² in 2015.

This is equivalent to less than two percent of the solar radiation reaching the Earth's surface, and the continuing increase of greenhouse gas levels could – depending on the rate of future fossil fuel consumption, generation of other greenhouse gases, and complex feedbacks between the atmosphere, oceans, and the biosphere – raise the average global surface temperature by anywhere between less than 2° and more than 5°C above today's mean by the year 2100, a faster rise than any experienced during the last million years of the biosphere's evolution. Many possible environmental, health, economic, and political consequences of this potential rise have been investigated most exhaustively by the Intergovernmental Panel for Climatic Change and summarized in its periodic reports starting in 1990 (the fifth assessment report was published in 2014).

Higher atmospheric CO_2 levels would improve water use efficiency in agriculture, boost yields of many common crops, and make some climates more livable; as the impact will be spatially uneven (warming will be much more pronounced in higher latitudes) some countries or regions may see overall net benefits, but others may be seriously affected. Higher average temperatures could increase both the intensity and the duration of extreme heat spells (with attendant health and fire effects and higher needs for air-conditioning) and, in the long run, could lead to substantial poleward shifts of ecosystem boundaries (forests would be able to grow in higher latitudes) and some tropical diseases. An accelerated water cycle would result in an overall increase of precipitation, but some areas would become drier while others would receive more irregular rains. The melting of mountain glaciers would affect stream flow, water supply, and hydroelectric generation, and even a modest rise of the mean ocean level (mostly due to the thermal expansion of water and amounting to less than fifty centimeters by 2100) would imperil low-lying coastal areas and would seriously affect the habitability of many Pacific islands.

There is little hope that future increases of greenhouse gas emissions will be significantly reduced by voluntary cuts in national energy consumption or widespread increases in conversion efficiency. Even if all the national pledges submitted at the Paris global warming conference in 2015 were completely fulfilled they would still lead to a further substantial increase of emissions and would fail to keep the global average temperature to below 2°C above pre-industrial levels.

As a result, there has been a growing interest in techniques of CO_2 sequestration, and a few experimental projects have demonstrated the feasibility of some methods, particularly pumping the gas into abandoned oil or gas reservoirs. But we must remember that the global combustion of fossil fuels currently emits more than thirty-two billion tonnes of the gas, and that only a small share of this vast amount comes from very large sources such as coal-fired electricity-generating plants and refineries that are conveniently located near hydrocarbon fields, abandoned deep coal mines, or other deep underground structures suitable for CO_2 disposal. Logistics, and the cost of sequestering even only half the generated CO_2 (mobile sources are obviously beyond the reach of this option) are mind-boggling.

A much less realistic sequestration option would be to seed ocean waters with iron, to stimulate phytoplankton production and so remove the atmospheric carbon as the dead phytomass sinks to the ocean bottom. But this is not a practical long-term solution. The undoubted fertilizing effect of iron enrichment is necessarily short-lived, as the dispersed iron rapidly sinks below the reach of light into the dark oceanic abyss, where no photosynthesis is possible. Moreover, fleets of ships would have to crisscross the ocean to distribute the micronutrients, and in many parts of the world ocean it is nitrogen or phosphorus, rather than iron, which limits phytoplanktonic photosynthesis. Equally questionable are other proposals for geoengineering solutions, including those changing the planet's radiation balance.

5
Energy in everyday life: from eating to emailing

Eating is the only obligatory energy input in our lives. Some people's demand for food barely goes above the essential minimum, made up of their BMR and the small amount of energy needed for maintaining personal hygiene: housebound elderly people and meditating Indian *sādhus* are excellent examples. The rest of us prefer to go well beyond just covering basal metabolic and indispensable activity demands. As soon as people move beyond subsistence living (when they eat just enough to carry out necessary tasks but their food supply has no significant surplus, so they can find themselves repeatedly undernourished), they diversify and transform their diets in ways that follow some remarkably consistent trends. I will describe these transformations, as well as some of their consequences, in the first section of this brief survey of energy in everyday life.

The second section will look at energy in our homes: how we keep warm (and cool) by burning fuels, and how we use electricity to extend the day, energize household appliances, and power the still-expanding selection of electronic devices whose uses range from heating food to storing and reproducing music. Globally, household energy use is claiming an increasingly higher share of overall energy consumption, and hence it is most welcome that almost all these diverse conversions have steadily improving efficiencies. I will detail several of these notable achievements, and then end the section with a brief look at the electricity demand of new electronic devices.

Thirdly, I will consider the energy needs of modern modes of transport. Increasing mobility is one of the most obvious attributes of modern civilization, and few factors correlate better with the overall standard of living than car ownership. In affluent countries, car ownership has long ceased to be a privilege, and many low-income countries are now acquiring cars at much faster rates than early adopters did at a similar stage of their economic development. The energy use of passenger cars thus deserves special attention – but the transformation of flying from an uncommon experience to everyday reality has been an even more revolutionary development. Flying has broadened personal horizons far beyond national confines: almost every airport able to receive large commercial airliners can be now reached from any other such airport in less than twenty-four hours flying time, many directly, others with changes of aircraft (the door-to-door time may be considerably longer, due to travelling time to and from the airport and infrequent connections). Remarkably, intercontinental travel, at close to the speed of sound, is an activity whose energy costs compare very favorably with many modes of contemporary driving.

The next section will address everyday energy encounters and realities. Many experiences exemplify large and incessant flows of energy, from the spectacular to the mundane, such as watching a television broadcast of a rocket launch from Cape Canaveral, seeing multi-lane traffic streaming down urban freeways, or walking past brightly-lit houses. On the other hand, few people think about energy flows when buying a plastic hamper, discarding a piece of aluminum foil, or installing a new staircase. But the objects around us do not have only mass, distinct shapes, function, or sentimental value: their production required fuel and electricity conversions, and each of them thus embodies a certain amount of energy. If you are a careful shopper you will know many of their prices: I will introduce you to some of their energy costs.

Finally, there is yet another hidden, or at least largely ignored, reality behind everyday energy use: it is increasingly probable that whenever you heat a house, start a car, or fly on vacation, you will rely on fuels that came not only from another country, but from another continent. Similarly, when you flip a light switch, play a Mozart concerto, or send an email it is increasingly likely the electrons performing those tasks originated in another country, or that your electricity was generated from imported fuels. We do not have to import toasters or toys from Asia, but everyday life for most of the world's people could not go on without intricate, and increasing, international dependence on traded energies.

Food intakes: constants and transitions

The fundamentals of human energetics are constrained by the necessities of heterotrophic metabolism: we have to cover our BMR just to be alive, we need additional food energy for growth and tissue repair, and we cannot lead healthy lives without either labor-related or recreational activities which, depending on duration and intensity, may require either marginal additions or substantial increments to our normal diet. Nearly everything else has changed, not only in comparison with traditional, pre-industrial, and largely rural settings, but also with early generations of the modern, industrialized, and urban world. Dietary transitions have profoundly changed the composition of average diets, mechanization and the use of agricultural chemicals have intensified food production, and socio-economic changes and food processing aimed at mass markets have introduced new eating habits.

Dietary transitions happen in all populations, as they become modern, industrial, and post-industrial urban societies (Figure 24).

These shifts share common features; national food balance sheets (reflecting food availability at retail level) indicate how far they have gone. Generally, as incomes rise so does the average per caput availability of food. The daily average is less than 2,000 kcal in the world's most malnourished populations, around 2,500 kcal in societies with no, or only a tiny, food safety margin, and well above 3,000 kcal in Europe, North America, and Australia. Remarkably, Europe's top rates, more than 3,300 kcal a day, are found not only in such rich countries as Germany, France Denmark and Belgium but also in Greece. The only notable departure from this high average is Japan, with about 2,700 kcal/day, now more than ten percent below China's mean!

Finding out how much food actually is consumed is a challenging task, and neither dietary recalls nor household expenditure surveys yield accurate results. To test the reliability of the first method, try to list every item (and the approximate amount) that you have eaten during the past three days and then convert these quantities into fairly accurate energy values! Unless itemized in great detail, a family's food expenditures tell us nothing about the actual composition (and hence overall energy content) of the purchased food, the level of kitchen waste, or consumption in a family.

The best available evidence, from a variety of food consumption surveys, shows that in affluent countries about 2,000 kcal are actually eaten per day per caput, with the average as low as 1,700 kcal/day for adult females and about 2,500 kcal/day for adult men. If these figures are correct, this means these countries waste 1,000–1,600 kcal per caput, or as much as a third to two-fifths of their available supply of food energy every day – and several national studies of food waste have confirmed this astounding level of loss. But some reported intakes underestimate real consumption: the extraordinarily high, and rising, obesity rates (particularly in North America, where about a third of people are obese and another third overweight) are

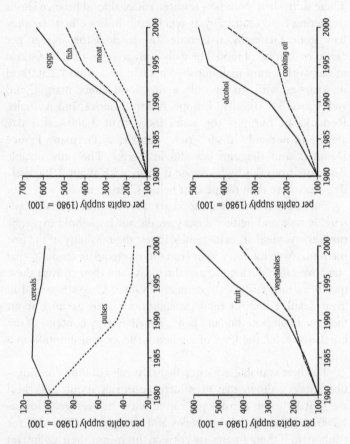

Figure 24 China's dietary transition, 1980–2000 (plotted from data in various editions of *China Statistical Yearbook*)

a result of insufficient physical activity and excessive food consumption. The second key feature of dietary transitions is a major shift in both relative and absolute contributions of basic macronutrients.

DIETARY TRANSITIONS

The most far-reaching dietary change has been the universal retreat of carbohydrate staples such as cereal grains (including rice, wheat, corn, and millet) and tubers (including white and sweet potatoes, and cassava). In affluent countries, they now supply just 20–30 percent of the average per caput energy intake, a third or half the traditional level. In Europe, this trend is shown by the declining consumption of bread, the continent's ancient staple: for example, in France, daily per caput intake fell from 600 g a day in 1880 to just 130 g a day by 2015, a nearly eighty percent drop. In Asia, rice consumption in Japan more than halved in the two post-World War II generations (to less than 60 kg annually per caput by 2000), making it an optional food item rather than a staple. A similarly rapid decline of rice intake has occurred in Taiwan, South Korea, and since the mid-1980s, China. This quantitative decline has been accompanied by a qualitative shift in cereal consumption, from whole grains to highly milled products (white flour). Post-1950 consumption of tubers also fell in all affluent countries, often by 50–70 percent compared to pre-World War II levels.

The retreat of starchy staples has been accompanied by a pronounced decline in the eating of high-protein leguminous grains (beans, peas, chickpeas, lentils, and soybeans). These were a part of traditional diets because of their unusually high protein content (twenty to twenty-five percent for most and about forty percent for soybeans), and every traditional society consumed them in combination with starchy staples (containing mostly just two to ten percent protein) to obtain essential dietary amino acids. As animal proteins became more affordable, traditional legume intakes tumbled to just a few kilograms per caput annually in North America and Japan. Among the more populous countries Brazil, at more than 15 kg a year (mostly black beans, feijão preto), has the highest per caput consumption of dietary legumes. Yet another key shift in carbohydrate consumption is the rising intake of refined sugar (sucrose), which was almost unknown in traditional societies, where sweetness came from fruits and honey. In some Western countries, the intake of added sugar exceeds 60 kg/caput annually (or up to twenty percent of all food energy), mainly because of excessively sweet carbonated beverages, confectionary and baked products, and ice cream.

The energy gap created by falling intakes of starchy foods has largely been filled by higher consumption of lipids and animal

---

DIETARY TRANSITIONS (*cont.*)

protein. Modern diets contain many more plant oils – both poly-unsaturated (peanut, rapeseed, and corn) and monounsaturated (olive) varieties, and saturated (coconut and palm) kinds – than did traditional intakes. Meat, animal fats, fish, eggs, and dairy products provided no more than ten percent of food energy in many traditional societies, but they now account for about thirty percent in affluent countries. The eating of meat has changed from an occasional treat to an everyday consumption, adding up annually to as much as 120 kg/caput (bone-in weight) in the U.S. and nearly 100 kg/caput in Spain. High-protein meat and dairy diets (with as much as sixty percent of all protein coming from animal foods) have resulted in substantial increases in average height and weight, and reoriented the rich world's agricultures from food crop to animal feed crop production.

Dietary transformations have changed some traditional food habits radically, and the Mediterranean diet perhaps best illustrates this. For decades, this diet has been extolled as the epitome of healthy eating with great cardiovascular benefits, and the main reason for the relatively high longevities in the region. But during the two post-World War II generations, there was a gradual change to increased intake of meat, fish, butter, and cheese, and decreased consumption of bread, fruit, potatoes, and olive oil. For example, olive oil now provides less than half of all lipids in Italy and Spain, and Spaniards now eat more meat than the Germans or French. The true Mediterranean diet now survives only among elderly people in rural areas.

Household energies: heat, light, motion, electronics

A modern house is a structure that provides shelter, but also contains a growing array of energy-conversion devices that increase the comfort, ease the daily chores, and provide information and entertainment for its inhabitants. In colder climates, heating usually accounts for most household energy; after World War II both Europe and North America saw large-scale shifts to more convenient, and more efficient, forms. Well-designed solid fuel (coal, wood, or multi-fuel) stoves have efficiencies in excess of thirty percent but, much like their wasteful traditional predecessors, they still require the laborious activities of bringing the fuel, preparing the kindling, starting the fire, tending it, and disposing of the ashes. Heating with fuel oil is thus a big advance in convenience (the fuel is pumped from a delivery truck into a storage tank and flows as needed into a burner) but it has been supplanted by natural gas.

For decades, affordability or aesthetics guided house design; energy consumption only became an important factor after OPEC's first round of price increases in 1973–1974. Cumulatively impressive savings in energy consumption can come from passive solar design (orienting large windows toward the southwest to let in the low winter sun), superinsulating walls and ceilings, and installing at least double-glazed, and in cold climates triple-glazed, windows. Fiberglass batting has an insulating value about eleven times higher than the equivalent air space, and more than three times higher than a brick. Consequently, the walls of a North American house framed with 4" x 6" wooden studs (2" x 4" are standard), filled with pink fiberglass, covered on the inside with gypsum sheets (drywall) and on the outside by wooden sheathing and stucco will have an insulation value about four times higher than a more sturdy looking, 10 cm thick, brick and stucco European wall. A triple-glazed window with a low-emittance coating (which keeps ultraviolet radiation inside the house), has an insulating value nearly four times as high as a single pane of glass. In hot climates, where dark roofs may get up to 50°C warmer than the air temperature, having a highly reflective roof (painted white or made of light-colored materials), which will be just 10°C warmer, is the best passive way to reduce the electricity needed for air conditioning – by as much as fifty percent. Creating a better microclimate, for example by planting trees around a house (and thus creating evapotranspirative cooling), is another effective way to moderate summer energy needs.

MODERN INDOOR HEATING AND COOLING

In North America, domestic furnaces heat air, which is then forced through metal ducts by an electric motor-powered fan and rises from floor registers (usually two to four per room). The best natural gas-fired furnaces are now about ninety-seven percent efficient, and hence houses equipped with them do not need chimneys. In Europe hot-water systems (using fuel oil and natural gas to heat water, which circulates through radiators) predominate.

Many Americans insist on raising their thermostats to levels that would trigger their air conditioning in summer (about 25°C); in most countries, the desirable indoor temperature is between 18 and 21°C. Even after their industrial achievements became the paragon of technical modernization, many Japanese did not have central heating in their homes; families congregated around the *kotatsu*, a sunken space containing, traditionally, a charcoal brazier, later replaced by kerosene, and then electric, heaters.

Engineers express the annual requirement for heating as the total number of heating-degree days: one day accrues for each degree that the average outdoor daily temperature falls below the specified indoor level. American calculations, based on a room temperature of 20°C, show that the coldest state, North Dakota, has about 2.6 times as many as does the warmest, Florida, while the Canadian values, based on 18°C, show Vancouver has less than 3,000 and Winnipeg nearly 6,000 heating-degree days a year.

Space cooling (first as single-room window units, later as central systems) began its northward march up North America as electricity became more affordable. Its spread changed the pattern of peak electricity consumption: previously the peaks were during the coldest and darkest winter months, but the widespread adoption of air conditioning moved short-term (hours to a couple of weeks) consumption peaks to July and August. Air conditioning is still relatively rare in Europe, but has spread to not only all fairly affluent tropical and subtropical places (Singapore,

MODERN INDOOR HEATING AND COOLING (*cont.*)

Malaysia, Brunei, Taiwan) but also to the urban middle class of monsoonal Asia (from Pakistan to the Philippines, with the largest concentrations in China's megacities) and humid Latin America. The relative costs of heating and cooling depend, obviously, on the prevailing climate and the desired indoor temperature.

But what makes modern houses so distinct from their predecessors is the still-expanding array of electricity uses, which requires an elaborate distribution network to ensure reliable and safe supply, and could not work without transformers.

Electricity is generated at voltages between 12.5 and 25 kV, but (as explained in Chapter 1) a combination of low current and high voltage is much preferable for long-distance transmission. So, the generated low current is first transformed (stepped-up) to between 138 and 765 kV before being sent to distant markets, transformed again (stepped-down) to safer, lower, voltages for distribution within cities (usually 12 kV) and then stepped-down to even lower voltages (110–250V depending on the country) for household use. Transformers can reduce or increase the voltage with almost no loss of energy, very reliably, and very quietly. They use the principle of electromagnetic induction: a loop of wire carrying an alternating current (the transformer's primary winding) generates a fluctuating magnetic field, which induces a voltage in another loop (the secondary winding) placed in the field, and vice versa. The total voltage induced in a loop is proportional to the number of its turns: if the secondary has twice as many turns as the primary, the voltage will double. Transformers range from massive devices with large cooling fins to small, bucket-size units mounted on poles in front of houses.

Household electricity use began during the 1880s with low-power lighting, and by 1920 had extended to a small selection of kitchen appliances. Refrigerators and radios came next, then, post-1950 Western affluence brought a remarkable array of electrically powered devices for use in the kitchen and workshop, and in recreation and entertainment. However, no other segment of modern energy use has seen such improvements in terms of efficiency, cost, and hence affordability, as has electric lighting. During the early period of household electrification, the norm was one low-power incandescent light bulb (40 or 60 W) per room, so a typical household had no more than about 200–300 W of lighting. Today, American houses will have commonly 60-80 lights (some in groups of two to six), adding up to more than 3,000 W. But it would be wrong to conclude that this house receives ten times as much light as a house in 1900: the actual figure is much higher and, moreover, the flood of new light is astonishingly cheaper.

The first, Edisonian, carbon-filament light bulbs of the early 1880s converted a mere 0.15 percent of electricity into visible radiation; even two decades of improved design later, they were still only about 0.6 percent efficient. The introduction of tungsten, the first practical metallic filament and placing it in a vacuum within the bulb, raised the performance to 1.5 percent by 1910; filling the bulb with a mixture of nitrogen and argon brought the efficiency of common light bulbs to about 1.8 percent by 1913. The development of incandescent lights has been highly conservative; hence the filament light bulb you buy today is essentially the same as it was four generations ago. Despite their inefficiency and fragility, and the fact that better alternatives became widely available after World War II, incandescent light bulbs dominated the North American lighting market until the end of the twentieth century.

The origin of more efficient, and hence less expensive light sources, predates World War I but discharge lamps entered the retail market only during the 1930s. Low-pressure sodium lamps came first, in 1932, followed by low-pressure mercury vapor lamps, generally known as fluorescent lights. These operate on an entirely different principle from incandescent lights. Fluorescent lights are filled with low-pressure mercury vapor and their inside surfaces coated with phosphorous compounds; the electrical excitation of the mercury vapor generates ultra-violet rays, which are absorbed by the phosphors and re-radiated in wavelengths that approximate to daylight. Today's best indoor fluorescent lights convert about fifteen percent of electricity into visible radiation, more than three times as much as the best incandescent lights, and they also last about twenty-five times longer (Figure 25).

Metal halide lights, introduced in the early 1960s, have a warmer color than the characteristic blue-green (cool) of early fluorescents, and are about ten percent more efficient. Another important step was to make the discharge lights in compact sizes, and with standard fittings, so they would not require special fixtures and could replace incandescent lights in all kinds of household applications. Initially, these compact fluorescents were rather expensive but large-scale production has lowered prices. Instead of a 100W incandescent light, all we need is a 23W compact, which will last 10,000 hours (nearly fourteen months

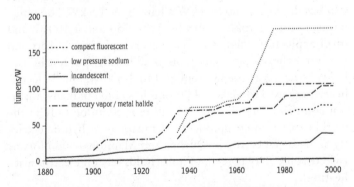

Figure 25 Improvements of lamp efficacy, 1880–2000 (based on an image at http://americanhistory.si.edu/)

of continuous light). Light emitting diodes (LED) are even better: they will last 50,000 hours and their efficacy (light produced per watt) will soon surpass that of today's most efficient low-pressure sodium lamps used for outdoor illumination.

When these technical advances are combined with lower electricity prices and higher real wages, electric light appears to be stunningly cheap. In the U.S., the average (inflation-adjusted to 2000 values) cost of electricity fell from 325 cents per kilowatt hour (kWh) in 1900, to about seven cents in 2015, while the average (inflation-adjusted) hourly manufacturing wage rose from $4 in 1900, to about $22 in 2015. Factoring in efficiency improvements, a lumen of U.S. electric light was three orders of magnitude (roughly 3,000 times) more affordable in 2015 for a family of a factory worker than it was in 1900! Only the post-1970 fall in microprocessor prices offers a more stunning example of performance and affordability gains.

The second most common category of household devices is appliances that pass current through high-resistance wires to generate heat. The highest demand comes from electric stoves (with ovens and usually four stovetop heating elements, adding up to as much as 8-9 kW on the highest settings). Clothes dryers rate up to 4 kW, a hair dryer 1.5 kW. The ubiquitous two-slice toaster rates between 750 and 1,100 W, and small appliances, such as coffee makers and rice cookers and hot water thermopots (common in all but the poorest households of East Asia), draw from 500 to 1,500 W. Small electric heaters are used to make cold rooms less chilly in many northern countries, but all-electric heating is common only in the world's two leading producers of inexpensive hydroelectricity, Canada (especially in Quebec and Manitoba) and Norway (where some sixty percent of households rely on electric space heating).

Small motors are the third important class of common household electricity converter. During Canadian winters, the most important motor (rated at 400–600W) is the one that runs the blower that distributes the air heated by the natural gas furnace in the basement. But the one that comes on most often is a smaller, 100–200 W, motor that compresses the working fluid in a refrigerator. Small motors also convert electricity to the mechanical (rotary) power needed for many household tasks in the kitchen or a workshop previously performed by hand. The single-phase induction motor, patented by Nikola Tesla (1856–1943) in 1888 and distinguished by its characteristic squirrel-cage rotor, is the most common. These sturdy devices run for years without any maintenance, powering appliances from sharp-bladed food processors and dough mixers (between 250–800 W) to floor, desktop, and ceiling fans (100–400 W).

Finally, there is a major and diverse class of electricity converters: electronic devices. The ownership of TVs and microwave ovens is almost universal, not only in affluent countries but among the growing middle classes of Asia and Latin America. Personal computers, rapidly ascendant during the 1980s and 1990s, are now much less popular than lighter, portable laptops, tablets and smartphones notebooks, while the largest flat-screen televisions have grown to wall-size dimensions. These new electronic gadgets have small unit-power requirements: flat screen televisions draw less than 100W, the active central processing units of desktop computers around 100 W, monitors up to 150 W, a laptop about 50 W. This means that, for example, emailing nonstop for twelve hours will consume 0.6 kWh, or as much electricity as an average clothes dryer will use in ten minutes. But given the hundreds of millions of computers (and printers, fax machines, copiers, and scanners) now owned by households, and the energy needed by the internet's infrastructure (servers, routers, repeaters, amplifiers), this latest domestic electricity market adds up already to a noticeable amount of the total electricity demand in the world's richest countries, and the web traffic will further increase that demand. For example, by 2020 the U.S. data centers are expected to use fifty-three percent more electricity than in 2013 and that would require about twenty large (1 GW) stations operating with seventy percent capacity factor.

Electronic devices are the main reason why modern households use electricity even when everything is turned off. All remote-controlled devices (televisions, video recorders, audio systems), as well as security systems, telephone answering machines, fax machines, and garage door openers use, even when idle, small (sometimes more than ten watts, usually less than five) amounts of electricity. In America these phantom loads (vampire power) now commonly add up to 50 W per household, and nationwide, they consume more electricity than is used annually in Hong Kong or Singapore. But most of these losses could be drastically reduced by installing controls that limit the leakage to less than 0.1 W per device.

Transport energies: road vehicles and trains

As individual economies become more affluent, the average number of people per car is converging toward the saturation level of just above, and in some cases, even slightly less than, two. This growth in car ownership has been achieved thanks to the mass production of affordable designs of family cars. Ford's Model T (1908–1927, a total of about fifteen million vehicles) was the trendsetter and the Volkswagen (Beetle) its single most successful embodiment. This car's initial specifications were made by Adolf Hitler in 1933, and its production (in Germany between 1945 and 1977, then in Brazil, and until 2003 in Mexico) amounted to 21.5 million vehicles. France's contribution was the Renault 4 CV, Italy's the Fiat Topolino, and Britain's the Austin Seven. With spreading affluence came more powerful high-performance cars, larger family cars, and, starting in the U.S. during the 1980s, the ridiculously named sports utility vehicle (SUV: what sport is it to drive it to work or a shopping center?).

In 2015 the average number of people per passenger vehicle was about 1.5 in Italy, and 1.7 in Japan, France and the U.S.. China is now the world's largest car market but in 2015 there were still nearly eight people per car. The total number of vehicles worldwide (passenger cars, buses, vans, trucks) passed 1.25 billion in 2014. Recent annual passenger car sales have been in excess of seventy million units, with nearly two-thirds sold in North America and China. Volkswagen, Toyota, GM and Renault-Nissan were the largest carmakers in 2015, selling about half of all vehicles made worldwide. The typical car use varies widely, from less than 5,000 km a year in some crowded Asian cities, to just above 10,000 km in major European countries (a rate that has been fairly steady for a long time), to about 18,000 km in the U.S.. Unfortunately, motor vehicles are also responsible for 1.25 million accidental deaths every year (and more than ten times as many serious injuries), and motor vehicle emissions are the principal cause of the photochemical smog that now affects, seasonally or continuously, almost all mega-cities and their surrounding areas (an increasing incidence of respiratory problems and damage to crops are its major impacts).

Refinery statistics show the total global output of about 3.7 billion tonnes in 2015, including nearly one billion tonnes of gasoline and more than 1.3 billion tonnes of diesel fuel. Passenger cars are by far the largest consumers of gasoline but some of it is also used by small airplanes, and by tens of millions of boats, snowmobiles, lawnmowers, and other small motors, while most of the diesel fuel is consumed by trucks, trains, and ships.

The fundamentals of internal combustion engines, the world's most abundant mechanical prime movers, have not changed for more than a century: a cycle of four strokes (intake, compression, combustion, exhaust), the use of gasoline and sparking ignition, and the conversion of the reciprocating movement of pistons to the rotary motion of a crankshaft. But a steady stream of technical improvements has made modern engines and transmissions lighter, more durable, and more reliable, and vehicles more affordable. The widespread use of microprocessors (American cars now have as many as fifty, with a combined power greater than that of the processors in the 1969 Apollo 11 lunar landing module) to control automatic transmission, anti-lock brakes, catalytic converters, and airbags has made cars more reliable, and converted them into complex mechatronic machines.

Despite these advances, internal combustion engines remain rather inefficient prime movers and the overall process of converting the chemical energy of gasoline to the kinetic energy of a moving passenger car is extraordinarily wasteful. A modern gasoline-fueled, four-stroke engine, in good repair, will convert no more than twenty percent of its fuel into reciprocating motion: the rest is waste heat transferred to the exhaust (nearly forty percent), cooling water (a similar share), and heat generated by engine friction. In towns or cities, where most of the world's cars are used, at least five percent of the initial energy input is lost idling at red lights; in cities with heavy traffic this loss may be easily of the order of ten percent. Finally, an increasing share of energy is used by auxiliary functions, such as power steering and air conditioning. This means that no more than thirteen and as little as seven percent of the energy of the purchased gasoline will make it to the transmission, where friction losses will claim a further five percent, leaving just two to eight percent of the gasoline actually converted to the kinetic energy of the moving vehicle.

ENERGY EFFICIENCY OF CARS

Even an efficient compact car (around six liters/100 km) will need nearly 2 MJ of fuel per kilometer. With a single occupant its driver (weighing about 70 kg) will account for around five percent of the vehicle's total mass, so the specific energy need will be nearly 30 kJ/kg of body mass. An adult briskly walking the same distance will need 250 kJ, 3.5 kJ/kg, an order of magnitude lower. This difference is not surprising, when we consider the huge energy waste needed to propel the vehicle's large mass (this has been steadily creeping up even for small European cars, from about 0.8 t in 1970 to 1.4 t in 2015; the average weight of new U.S. vehicles is now 1.8 t).

What is surprising is that even the most efficient cars have an energy cost per p-km only marginally better than that of the latest airliners; for many vehicles the cost is much higher than flying. High fuel taxes in Europe and Japan have kept vehicle sizes smaller there than in North America, where the first deliberate improvements in efficiency were forced only by OPEC's two rounds of dramatic crude oil price increases. There, the Corporate Average Fuel Economy (CAFE) standards doubled the fleet's performance to 27.5 mpg (8.6 l/100 km) between 1973 and 1987: better engines and transmissions, lighter yet stronger bodies, and reduced aerodynamic drag were the main factors. Japanese automotive innovations were particularly important for making efficient, reliable, and affordable cars. In the 1990s, Toyotas (basic Tercels and upscale Camrys) and Hondas (Civics and Accords) became America's bestselling cars; by the end of the twentieth century, most of them were actually made in the U.S. and Canada. Unfortunately, the post-1985 decline of crude oil prices eliminated further CAFE gains, and the overall fleet performance actually worsened with the rising popularity of vans, pick-up trucks, and SUVs. CAFE standards began to rise again in 2011 (to 30.2 mpg) and the goal is 50 mpg by 2022. Even so, in 2016 the least efficient SUV (Mercedes Benz AMG G65) rated just 12 mpg (19.6 l/100 km) so that with a driver and a single passenger that vehicle consumes 3.4 MJ/p-km. In contrast, the first generation of airliners consumed about 5 MJ/p-km, the Boeing 777 needs about 1.5 MJ/p-km on a short flight and the Boeing 787 (Dreamliner) on an intercontinental flight will use as little as 0.9 MJ/p-km.

Standard fuel economy ratings are established during controlled tests, which simulate city and highway driving. Actual performance in everyday driving is rarely that good. What can we do to improve it? Under-inflated tires are probably the most common cause of easily avoidable energy loss: their flexing absorbs energy, and their larger footprint increases rolling friction and heats up the tire. Bad driving (and idling) habits (rapid acceleration, or running the engine when waiting for more than a minute or so) add to wasted fuel. As far as the choice of fuel is concerned, there is absolutely no advantage, in higher efficiency, cleanliness, or speed, in filling up with a premium (high-octane) gasoline. Octane ratings rise with the fuel's ability to resist knocking; all new cars are designed to run well on regular gasoline (87 octane). No matter what the make or type of vehicle, the efficiency of driving follows a pronounced hump-shaped curve: at its maximum between 45 and 55 km/h, somewhat worse (ten to twenty percent) at lower speeds, but as much as forty percent less efficient at speeds above 100 km/h. This, and the noticeable reduction in fatal accidents, is the key argument for limiting the maximum speed to no more than 110 km/h.

In contrast, the fastest rapid intercity trains now travel at up to 300 km/h, and each of them can carry more than 1,000 passengers in comfortable, airline-style seats. They have an accident rate nearly an order of magnitude lower than driving, and no form of land transport can match their low energy cost per p-km (nearly always below 1 MJ). The Japanese *shinkansen*, the first rapid train (its scheduled service started on October 1, 1964) and the French *TGV* move people with less than 0.4 MJ/p-km, their maximum short-term speeds in regular service are close to, or even above, 300 km/h.

All rapid trains are powered by electric motors supplied via pantographs from overhead copper (or copper-clad steel) wires. The *shinkansen* pioneered the design of a rapid train without a locomotive: instead, every car has four motors (the latest 700 series trains have sixty-four motors with a total capacity of 13.2 MW). This arrangement makes the frequent accelerations and decelerations of relatively short inter-station runs much easier, moreover, the motors act as dynamic brakes, when they become generators driven by the wheels. The best proof of the train's (and track's) admirable design and reliability is the fact that during the first fifty years of its service (1964-2014) the Tōkaidō line between Tōkyō and Ōsaka had carried more than five billion people without a single accidental fatality (Figure 26).

The *TGV* (its first line has been operating since 1981) also uses its synchronous motors for dynamic braking at high speeds but, unlike the *shinkansen*, it has two locomotives (each rated at 4.4 MW) in every trainset. Several European countries have been trying to catch up with French accomplishments: Spain got its

Figure 26 Tōkaidō line Nozomi train in Kyōtō station (photo Vaclav Smil)

first *TGV* train (the Madrid-Seville *AVE*) in 1991, while Italy's *ETR500* has been operating since 1993, both with top speeds of 300 km/h. In contrast, Amtrak's New York to Washington *Acela*, with a peak speed of 240 km/h, remains North America's only rapid train service, a singularity explained by low population density and highly competitive road and air links. But every large metropolitan area in North America, Europe, and Japan has several major lines, and many have fairly dense networks of slower commuter trains powered either by electricity or diesel locomotives.

Flying high: airplanes

Commercial flight remained a rare, expensive, and uncomfortable experience as long as it was dominated by propeller engines (that is, until the late 1950s). Passengers had to endure insufficiently pressurized cabins, propeller noise, vibration induced by the four-stroke engines, and a great deal of turbulence (low cruising altitudes subjected the relatively small airplanes to bumpy conditions), for many hours. The speed of propeller airplanes gradually increased, but even the best machines did not pass 320 km/h, and hence it took (with three stops needed for refueling) fifteen and a half hours to fly from New York to Los Angeles and more than eighty hours from England to Japan. The capacity of these planes was also limited: the legendary DC-3 (first flown in 1935) accommodated at most thirty-six people and PanAm's huge Clipper could take only seventy-four passengers.

Jet airplanes changed everything, with maximum air speeds above 900 km/h, capacities of up to five hundred people, trans-American crossing in less than six hours, and cruising altitudes of 10–12 km, high above tropospheric turbulence. The first modern jet airliners were derived from the largest post-World War II military planes: the most important was the B-47 Stratojet bomber, with its swept wings and engines hung on struts under the wings, two enduring design features of all large commercial jets. Many innovations were needed to improve the performance and reduce the operating cost of these machines to make air travel an affordable and fairly common event. Stronger aluminum alloys, entirely new composite materials, advances in aerodynamic design, and ingenious electronic navigation and landing devices (radar above all), were the essential ingredients but better engines made the greatest difference.

All jet airplanes are propelled forward as a reaction to the backward thrust of their engines. The first commercial planes, of the 1950s, used slightly modified military turbojet engines. These compact machines compressed all the air that passed through the frontal intake (compression ratios were just 5:1 for the earliest designs; now they commonly go above thirty) and fed it into the combustor, where a fine fuel spray was ignited and burned continuously; the escaping hot gases first rotated a turbine which drove the compressor and as they left through a nozzle, created the thrust that propelled the plane. Thrust reversers are flaps deployed after landing to force the gas forward and slow down the aircraft (the engine cannot reverse the flow of hot gas). Turbojets were also configured to rotate (by using reduction gears) propellers: these turbo-props are now common only among small commuter aircraft, while all large planes are now powered by turbofans.

TURBOFAN ENGINES

Turbojets have two basic disadvantages for commercial aviation: peak thrust occurs at very high, supersonic speeds and fuel consumption is relatively high. These drawbacks can be remedied by placing large-diameter fans in front of the engine, driven by a second turbine placed after the primary rotor that powers the compressor; they compress the incoming air to about twice the

TURBOFAN ENGINES (*cont.*)

inlet pressure. Because this compressed air bypasses the combustor, it reduces specific fuel consumption but adds another (cool and relatively slow) stream of exhaust gases to the rapid outflow of hot gas from the core and generates more thrust. Figure 27 is the cutaway view of GE90, a turbofan powering Boeing 777 jetliners, with a bypass ratio of nine. The highest bypass ratios in 2016 were Rolls-Royce Trent 1000 turbofan (used to power Boeing 787) at 10 and Pratt & Whitney's PW1000G (powering Airbus A320neo and Bombardier CSeries) at 12. Today's turbofans can get planes airborne in less than twenty seconds, guaranteeing a much less harrowing experience for white-knuckled fliers!

Turbofans are also much quieter, as the high-speed core exhaust is enveloped by slower-moving bypass air. The diameters of the largest turbofans are nearly 3.5m (or just a few percent less than the internal fuselage diameter of a Boeing 737!) and because the temperature of their combustion gases (around 1500°C) is above the melting point of the rotating blades an efficient internal air cooling is a must. Despite such extreme operating

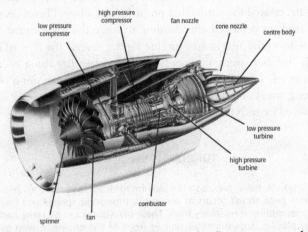

Figure 27 Cutaway view of GE90 jet engine (image courtesy of General Electric)

TURBOFAN ENGINES (*cont.*)

conditions, turbofans, when properly maintained, can operate for up to 20,000 hours, the equivalent of two years and three months of non-stop flight. Their extraordinary reliability results in fewer than two accidents per million departures, compared to more than fifty for the turbojets of the early 1960s.

With so much power at our command, it is not turbofans that now limit the size of commercial planes but rather the expense of developing new designs, unavoidable infra-structural constraints (length of runways, loading and offloading capacity, number of terminal slots), and safety considerations that make planes of more than 1,000 passengers unlikely. The twin-deck Airbus 380 has been certified for the maximum of 853 passengers but actual configurations in service range from 379 passengers (Singapore Airlines, three-class layout) to 615 passengers (Emirates, two-class layout). The development of the A380 was a key strategic move by Airbus (established in 1970) as it fights Boeing for dominance in the global market for large passenger aircraft. In 2003 Airbus was, for the first time, slightly ahead of Boeing in terms of overall annual deliveries for large passenger planes (but Boeing regained the primacy in 2012). Its most common planes are the smaller jets of the A320 family (107-230 passengers) and the larger planes of the A330 and A340 (four-engine) family, typically for about 350 passengers. But the Boeing 737 remains the bestselling airliner, and the pioneering, and daring 747 (Figure 28) remains the most revolutionary, if not the best, ever built (with more than 1,500 sold by 2016). Boeing's other highly successful designs include the 767, which has dominated trans-Atlantic flights, and the 777, whose model 200 LR (test-flown in March 2005), is the world's longest-range (17,395 km) airplane, able to fly directly between any two airports in the world. Boeing's latest model, 787, was designed to be the world's most energy-efficient

Figure 28 Boeing 747-8I in Lufthansa service (from Wikimedia)

and most comfortable airliner, with a new cabin layout, larger windows, and better air quality but its service entry was delayed by six years and its final cost was a multiple of the original budget projection.

Kerosene is a much better fuel for jet engines than gasoline: it has a higher specific density (0.8 as against 0.72 g/l) and hence higher energy density (34.5 compared to 32.0 MJ/l) so more can be stored in available tanks, and as a heavier refined fuel it is cheaper, it has lower evaporation losses at high altitudes, a lower risk of fire during ground handling, and produces more survivable crash fires. Jet A fuel, used in the U.S., has a maximum freezing point of −40°C, while Jet A-1, with a freezing point of −47°C, is used on most long international flights, particularly on northern and polar routes during winter. Airliners store their fuel in their wings; some also have a central (fuselage) and a horizontal stabilizer tank. They must carry enough fuel for the projected flight, and for any additional time that might be needed due to weather conditions (such as unexpectedly strong headwinds), or airport congestion.

As already noted, airliners offer a surprisingly energy-efficient form of long-distance passenger transport. Improvements of inherently highly efficient gas turbines, and their ability to carry fairly comfortably hundreds of passengers mean they waste less fuel per passenger than intercity two-passenger car drives longer than about 500 km. The unsurpassed rapidity and convenience of long-distance flying pushed the global figure to nearly six trillion p-km during 2016, a clear testimony of the extent to which jet-powered flight has changed the way we do business, maintain family ties, and spend our free time. And because of more efficient jet engines, fuel consumption has been rising at a somewhat slower rate and by 2015 it was equivalent to about seven percent of all liquid fuels used in transportation.

Embodied energies: energy cost of materials, food – and energy

People have always asked "how much" when buying goods and services but before OPEC's first round of steep oil price increases only those companies whose energy bills accounted for most, or a very large share, of their overall production costs traced their energy expenditures in detail, to better manage and reduce those outlays. With rising energy prices came many studies that calculated the energy costs of products ranging from bulk industrial goods (basic metals, building materials, and chemicals) to consumer products (cars to computers), to foodstuffs. This information has not changed the ways of an average Western consumer, who is still utterly unaware of either the overall or the relative energy cost of everyday products and services, but it has been very helpful to producers trying to reduce the energy cost of their activities: only a detailed account of the fuel and electricity needs of the individual processes or the components of a product can identify the management opportunities and technical improvements that will help to minimize those outlays.

Energy cost analysis is simple only when it is limited to a single process, with one, or a few, obvious direct energy inputs (for example, coke, natural gas and electricity for producing pig iron in a blast furnace). It becomes very complex, and involves debatable choices of analytical boundaries, when the goal is to find the overall energy cost of such complex industrial products as a passenger car. A complete analysis should account not only for energies used in the assembly of the car, but also for fuels and electricity embodied in structural materials (metals, plastics, rubber, glass), as well as in the vehicle's electronic controls. The third level of energy cost finds the costs of capital equipment, used to produce major material inputs and build the production facilities. The next step is to find the energy cost of the energy that has made all this possible, but at that point it becomes clear that the inclusion of higher-order inputs will result in rapidly diminishing additions to the overall total: capturing the first two or three levels of a product's energy cost will usually give us eighty to ninety percent of the total.

It should come as no surprise that among many published figures of energy cost there are nearly identical values, reflecting the widespread diffusion of numerous advanced industrial techniques, or the worldwide use of identical processes or products: as noted in the previous section, the world has now only two makers of large passenger airplanes, and only three companies make most of their engines. International comparisons show many basically identical products with substantially different energy inputs: such disparities are caused by differences in industrial infrastructure and management. I will comment on the energy costs of a small number of the most important items (judged by the overall magnitude of their output) in four key categories: bulk raw materials, large-volume industrial inputs, major consumer products, and basic foodstuffs. All values will be expressed in gigajoules per tonne (GJ/t) of the final product (for comparison, energy content of one tonne of crude oil is 42 GJ/t).

Basic construction materials require only extraction (wood, sand, and stone) and heat processing (bricks, cement, and glass). Excavating sand may take as little as 0.1 GJ/t, quarrying stone less than 1 GJ/t, despite the near-total mechanization of modern lumbering, a tonne of construction wood costs mostly between 1.5–3 GJ/t, and the best ways to produce cement now require less than 3.5 GJ/t. Making concrete (a mixture of cement, sandy or gravel aggregate, and water) costs very little additional energy, but reinforcing concrete with steel makes it nearly three times as energy expensive. The most energy-intensive materials commonly used in house construction are insulation and plate glass, up to 10 GJ/t. Integrating these inputs for entire buildings results in totals of around 500 GJ (an equivalent of about 12 tonnes of crude oil) for an average three-bedroom North American bungalow, and more than 1,500 times as much for a 100 storey skyscraper of 1000 m^2 per floor. In commercial buildings and residential highrises, steel is most commonly the material with the highest aggregate energy cost.

ENERGY COST OF METALS

Steel remains the structural foundation of modern civilization: it is all around us, exposed (in vehicles, train and ship bodies, appliances, bridges, factories, oil drilling rigs, and transmission towers) or hidden (in reinforced concrete, and skyscraper skeletons); touched many times a day (kitchen and eating utensils, surgical instruments, and industrial tools) or buried underground (pipes, pipelines, and piled foundations). Globally, about thirty percent is now made from recycled scrap, but most still comes from large blast furnaces. Technical advances lowered the typical energy cost of pig iron (so-called because it is cast into chunky ingots called pigs) to less than 60 GJ/t

ENERGY COST OF METALS (*cont.*)

by 1950 and to just 20 GJ/t by the year 2000. The most highly efficient operations now produce semi-finished iron products (ingots,
blooms, billets, slabs) for less than 20 GJ/t. Pig iron is an alloy that
contains from 2 to 4.3 percent carbon, while steel has commonly
just 0.3–0.6 percent and no more than one percent). This seemingly
small quantitative difference results in enormous qualitative gains:
cast iron has poor tensile strength, low impact resistance, and very
low ductility.

Steel has high tensile strength, high impact resistance, and
remains structurally intact at temperatures more than twice as
high as iron. Its alloys are indispensable, making everything
from stainless cutlery to giant drilling rigs. For nearly a century,
pig iron's high carbon content was lowered, and steel made by
blasting the molten metal with cold air in open hearth furnaces;
only after World War II were these replaced by basic oxygen
and electric arc furnaces. The processing of steel was concur-
rently revolutionized by abandoning the production of steel
ingots (which had to be reheated before they were shaped
into slabs, billets, or bars) and instead continuously casting the
hot metal. These innovations brought enormous (up to one
thousand fold) productivity gains, and great energy savings.
The classic sequence of blast furnace, open hearth furnace,
and steel ingots needed two to three times as much energy
for semi-finished products as the modern combination of blast
furnace, basic oxygen furnace, and continuous casting.

The reduction of non-ferrous metals from their ores requires
substantially more energy than iron-smelting. The production
of aluminum from bauxite remains, despite substantial efficiency gains during the twentieth century, particularly energy-
intensive, averaging about 175 GJt (titanium, widely used in
aerospace, needs four times as much). Hydrocarbon-based plastics have replaced many metallic parts in vehicles, machines, and
devices, because of their lower weight and resistance to corro-
sion, but their energy cost is fairly high, between 75–120 GJ/t.
Motor vehicles are the leading consumers of metals, plastics,
rubber (another synthetic product), and glass. Their energy cost
(including assembly) is typically around 100 GJ but this accounts
for no more than about twenty percent of the vehicle's lifetime
energy costs, which are dominated by fuel, repairs, and road
maintenance.

The energy costs of common foodstuffs range widely, due to different modes of production (such as intensity of fertilization and pesticide applications, use of rain-fed or irrigated cropping, or manual or mechanized harvesting) and the intensities of subsequent processing. The typical costs of harvested staples are about 4 GJ/t for wheat, corn, and temperate fruit, and around 10 GJ/t for rice. Produce grown in large greenhouses is most energy intensive; bell peppers and tomatoes cost as much as 40 GJ/t. Modern fishing has a similarly high fuel cost per tonne of catch. These rates can be translated into interesting output/input ratios: harvested wheat contains nearly four times as much energy as was used to produce it but the energy consumed in growing greenhouse tomatoes can be up to fifty times higher than their energy content.

These ratios show the degree to which modern agriculture has become dependent on external energy subsidies: as Howard Odum (1924-2002) put it in 1971, we now eat potatoes partly made of oil. But they cannot simplistically be interpreted as indicators of energy efficiency: we do not eat tomatoes for their energy but for their taste, and vitamin C and lycopene content, and we cannot (unlike some bacteria) eat diesel fuel. Moreover, in all affluent countries, food's total energy cost is dominated by processing, packaging, long-distance transport (often with cooling or refrigeration), retail, shopping trips, refrigeration, cooking, and washing of dishes: at least doubling, and in many cases tripling or quadrupling, the energy costs of agricultural production.

According to many techno-enthusiasts, advances in electronics were going to lead to the rapid emergence of a paper-free society, but the very opposite has been true. The post-1980 spread of personal computers has been accompanied by a higher demand for paper. Since the late 1930s, global papermaking has been dominated by the sulfate process, during which ground-up coniferous wood and sulfate are boiled, under pressure, for about four hours to yield a strong pulp that can either be used to make unbleached paper, or bleached and treated to produce higher quality stock. Unbleached packaging paper takes less than 20 GJ/t; standard writing and printing stock is at least forty percent more energy intensive (requiring more energy than steel!).

Last, but clearly not least, a few numbers regarding the energy cost of fossil fuels and electricity. These costs are usually expressed as energy return on investment (EROI). Given the wide range of coal qualities and great differences in underground and surface coal extraction (see Chapter 4), it is not surprising that coal's EROI ranges widely, between 10 (for underground coal in thinner seams) and 80 (for large surface mines). Similarly, crude oil from the richest Middle Eastern fields can contain far more than 100 times as much energy as it costs to produce it, but the EROI for small oil fields of low productivity can be no higher than 10. But refining requires more energy to separate the crude oil into many final products and it will raise the final energy costs of refined oil products.

As we have seen, thermal electricity generation in large power plants (with boilers and steam turbogenerators) is, at best, about forty percent efficient; typical rates (including high-efficiency flue gas desulfurization and the disposal of the resulting sulfate sludge) may be closer to thirty-five percent. The energy costs of constructing the stations and the transmission network equate to less than five percent, and long-distance transmission losses subtract at least another seven percent. Electricity produced in a large thermal station fueled by efficiently produced surface-mined bituminous coal would thus represent, at best, over thirty percent, and more likely from twenty to twenty-five percent of the energy originally contained in the burned fuel. Because of their higher construction costs, the net energy return is lower for nuclear stations, but we cannot do a complete calculation of the energy costs of nuclear electricity because no country has solved the potentially very costly problem of long-term disposal of radioactive wastes.

Global interdependence: energy linkages

In pre-industrial societies, the fuels needed for everyday activities overwhelmingly came from places very close to the settlement (for example, in villages, wood from nearby fuel wood lots, groves or forests, and crop residues from harvested fields) or were transported relatively short distances. There were some longer shipments of wood and charcoal, but international fuel deliveries became common only with the expansion of coal mining and the adoption of railways and steam-powered shipping. The energy transitions from coal to crude oil and natural gas, and the growing prominence of electricity have profoundly changed the pattern of energy supply − yet, who, during the course of daily activities, thinks about these impressively long links?

Electricity used by an English household may have been generated by burning coal brought from Colombia, and the coking coal used to produce Chinese pig iron could have come not only from North China but also from Australia and Canada; the gasoline in a New Yorker's car may have originated as crude oil pumped from under the ocean floor in the Gulf of Mexico, some 2,000 km away, refined in Texas and taken by a coastal tanker to New Jersey; the natural gas used to cook rice in a Tokyo home may have come by tanker from Qatar, a shipping distance of nearly 15,000 km; and the electricity used to illuminate a German home may have originated as falling water in one of Norway's hydroelectricity stations.

Energy accounts for a growing share of the global value of international trade: about eight percent in 2000, seventeen percent by 2014. The global fuel trade added up to more than $3 trillion in 2014 (twice as much as the trade in all agricultural products). In mass terms, the global fuel trade – in 2015 more than a trillion cubic meters of natural gas, more than 1.2 billion tonnes of coal, and nearly three billion tonnes of crude oil and refined products – towers above all other bulk commodities, such as ores, finished metals, and agricultural products. Crude oil leads both in annually shipped mass and monetary value. In 2015 the Middle East accounted for a third of all crude oil exports, and the EU, China, U.S. and Japan for nearly 60 percent of all crude oil imports.

Tankers carry about two-thirds of all crude oil and refined products exports from large terminals in the Middle East (Saudi's Rās Tanūra is the world's largest), Africa, Russia, U.S., Latin America, and Southeast Asia to huge storage and refining facilities in importing countries (U.S., a large crude oil importer, is also a large exporter of refined products). The rest of the world's crude oil exports go by pipelines, the safest and cheapest form of mass land transport. The U.S. had constructed an increasingly dense network of oil pipelines by the middle of the twentieth century, but major export lines were only built after 1950. The longest (over 4,600 km, with an annual capacity of 90 million tonnes) was laid during the 1970s to carry oil from Samotlor, a super-giant field in Western Siberia, first to European Russia and then to Western Europe.

In contrast to crude oil, only thirty percent of the world's natural gas production was exported in 2015. Two-thirds of it is moved through pipelines. Russia, Canada, Norway, the Netherlands, and Algeria are the largest exporters of piped gas; the U.S., Germany, and Italy its largest importers. The world's longest (6,500 km) and largest-diameter (up to 1.4 m) pipelines carry gas from West Siberia's super-giant fields to Italy and Germany. There they meet the gas networks that extend from Groningen, the giant Dutch field, the North Sea field (brought first by undersea pipelines to Scotland), and Algeria (crossing the Sicilian Channel from Tunisia and then the Messina Strait to Italy).

Overseas shipments of natural gas became possible with the introduction of liquefied natural gas (LNG) tankers, which carry the gas at −162°C in insulated steel tanks; on arrival at their destination the liquefied cargo is re-gasified and sent through pipelines. The first LNG shipments were sent from Algeria to France and the U.K. during the 1960s; in 2015 LNG accounted for a third of natural gas exports. The major importers are Japan (which buys a third of world's supply, from the Middle East, Southeast Asia, and Alaska), South Korea, China and Taiwan.

In comparison to large-scale flows of fossil fuels, the international trade in electricity is significant in only a limited number of sales or multinational exchanges. The most notable one-way transmission schemes are those connecting large hydrogenerating stations with distant load centers. Canada is the world leader, selling hydroelectricity from British Columbia to the Pacific Northwest, from Manitoba to Minnesota, the Dakotas and Nebraska, and from Québec to New York and the New England states. Other notable international sales of hydroelectricity take place between Venezuela and Brazil, Paraguay and Brazil, and Mozambique and South Africa. Most European countries participate in a complex trade in electricity, taking advantage of seasonally high hydrogenerating capacities in the Scandinavian and Alpine nations.

6
Energy in the future: trends and unknowns

This closing chapter offers no forecasts; there is no need to add to the large, and growing, volume of that highly perishable commodity. Reviews show that most long range (more than ten to fifteen years ahead) energy forecasts – whether at sectoral, national, or global level, and no matter if they were concerned with the progress of individual techniques, the efficiency gains of a particular process, overall energy demand and supply, or the price levels of key commodities – tend to fail in a matter of years, sometimes months. Given the post-World War II penchant for long-range forecasting, it is now possible to recite scores of such failures. Perhaps the most tiresomely notorious is the ever-elusive further fifty years that will be needed to achieve commercial nuclear fusion (generating electricity by fusing the nuclei of the lightest elements – the same kind of reactions that power the Sun). Common failures include forecasts of the imminent global peak oil production, and some of the most spectacular misses include the predictions of future crude oil prices (too high or too low, never able to catch the reality of highly erratic fluctuations).

Even if some individual numbers come very close to the actual performance, what is always missing is the entirely new context in which these quantities appear. Imagine that in 1985 (after the collapse of crude oil prices and a sharp drop in global oil production), you accurately forecast global oil production in 2015. Could anybody in 1985 have predicted the four events that changed the post-1990 world: the peaceful collapse of the U.S.S.R. (first leading to a rapid decline and then to an impressive resurgence of its oil output); the emergence of China as the world's largest (in terms of purchasing power parity) economy (and also the world's largest importer of oil); September 11, 2001 (with its manifold consequences and implications for the world in general, and the Middle East in particular); and the worst post-World War II global economic downturn in 2008-2009?

No forecasts then – only brief reviews of some key factors that will determine the world's future quest for a reliable and affordable energy supply, and the major resource and technical options we can use during the next half-century. During that time, the basic nature of global energy supply will not drastically change, and the world will remain highly dependent on fossil fuels. At the same time, we know our fossil-fueled civilization to be a relatively short-lived phenomenon, and the next fifty years will see an appreciable shift toward non-fossil energy resources. At the beginning of the twentieth century, the world derived about sixty percent of its energy from coal, crude oil, and (a very little) natural gas. In 2015 eighty-two percent of the world's total primary energy came from fossil fuels, with the rest supplied by primary electricity (hydro and nuclear) and phytomass fuels.

Even if the recoverable resources of fossil fuels (particularly those of crude oil and natural gas) were much larger than today's best appraisals, it is clear they are not large enough to be the dominant suppliers of energy for an affluent civilization for more than a few centuries. Conversely, the combination of rapidly rising demand and the escalating costs of fuel extraction may limit the fossil fuel era to the past and present centuries – and the rapid progress of pronounced global warming, clearly tied to the combustion of fossil fuels, may force us to accelerate the transition to non-fossil energies. As already stressed in Chapter 1, the overall magnitude of renewable energy flows is not a constraint.

Biomass energies have been with us ever since we mastered the use of fire: wood, charcoal, crop residues, and dung are still used by hundreds of millions of peasants and poor urban residents in Asia, Latin America, and particularly throughout sub-Saharan Africa, mostly for cooking and heating. Our best estimates (there are no reliable statistics, as most of these fuels are collected or cut by the users themselves) put the worldwide energy content, at the beginning of the twenty-first century, of traditional biomass energies at about 40EJ, roughly eight percent of the world's aggregate primary energy consumption. But the share is much lower when comparing useful energies, because most of the biomass is burned very inefficiently in primitive stoves. As already noted in Chapter 3, these wasteful uses also have considerable health costs, due to indoor air pollution, and there are also the serious environmental problems of deforestation and the reduced recycling of organic matter. Biomass energies could make a difference only when harnessed by modern, highly efficient techniques without serious environmental and social impacts: achieving this will be an enormous challenge.

Hydroenergy is the only kind of indirect solar energy flow extensively exploited by modern techniques, but, outside Europe, North America, and Australia, there is still considerable untapped potential – but also a growing opposition to environmental impacts of large hydro projects. As already emphasized, potentially the most rewarding, and by far the largest, renewable energy resource is the direct solar radiation that brings close to 170W/m^2 to the Earth. Post-2000 development of PV electricity generation has been growing exponentially, as has the installation of wind turbines. There is also the possibility of new designs of inherently safer and more economic nuclear electricity generation. I will review the advantages and drawbacks of all of these major non-fossil options. But before doing so I must stress the magnitude of future energy needs against the background of enormous consumption disparities and long-term energy transitions.

Energy needs: disparities, transitions, and constraints

The extent of future global energy needs cannot be understood without realizing the extent of existing consumption disparities. The per caput annual energy consumption in the U.S. and Canada is roughly twice as high as in Europe or Japan, more than three times as high as in China, more than fifteen times as high as in India, and more than thirty times as high as in the poorest countries of sub-Saharan Africa. Because of this highly skewed consumption pattern, the global annual average of about 75 GJ is largely irrelevant: global distribution of average consumption rates is bi-modal, with most of the rates for low-income countries below 30 GJ, and with high-income countries average above 150 GJ.

UNEQUAL ACCESS TO MODERN ENERGY

The enormous disparity in access to energy is most impressively conveyed by contrasting continental shares of the global population with their corresponding share of worldwide primary energy consumption: Africa, with about seventeen percent of the global population, consumed about four percent of all primary commercial energy in 2015; North America and Europe, with a nearly identical combined shares of global population, consumed almost forty percent. The most stunning contrast: the U.S. alone, with less than five percent of the world's population, claims seventeen percent of the world's primary commercial energy.

No indicator of high quality of life – very low infant mortality, long average life expectancy, plentiful food, good housing, or ready access to all levels of education – shows a substantial gain once the average per caput energy consumption goes above about

MAGNITUDES OF ENERGY AND POWER (*cont.*)

100 GJt/year. Consequently, it would be rational to conclude that the world's affluent nations have no need to increase their already very high averages. At the same time, there are still hundreds of millions of people in the poorest countries who do not directly consume any fossil fuels and whose energy consumption must rise (Figure 29).

Because almost all the world's population growth during the first half of the twenty-first century will take place in low- and medium-income countries (affluent populations, with the exception of the U.S., will either be stagnant or in decline), most future increases in fossil fuel and electricity consumption will be in Asia and in Africa. But there is no easy way to forecast this new demand, as it is a complex function of population growth and economic expansion, the changing composition of the primary energy supply and final energy uses (energy-intensive heavy industries compared to light manufacturing and service industries), and the adoption of new inventions and more efficient energy converters.

Figure 29 Where people need more energy: conditions in this Jakarta slum are replicated in scores of low-income countries (reproduced from Wikipedia)

To achieve a modicum of economic security, the average annual per caput consumption rates should more than triple in sub-Saharan Africa, and more than double in India. It is clear that future energy use in the world's most populous and rapidly expanding economies will conform (with variations for national characteristics) to the general pattern of energy transitions that has taken place in affluent countries. Their two principal components were noted in Chapter 4: the declining share of coal in total primary consumption (although in China that share will remain relatively high, because of its extraordinary dependence on the fuel), and a steady rise of oil and natural gas consumption, leading to higher demand for imported liquids and gases.

Another key ingredient of the worldwide transition of commercial energy use has been the rising share of electricity in final consumption. In 1900, less than a generation after the beginning of electricity generation, little more than one percent of fossil fuel consumption was converted to electricity; by 1950 the global share rose to ten percent and by 2015 it reached thirty percent. Nearly everywhere, electricity use has been growing at a much faster rate than the consumption of fuels, because during the second half of the twentieth century fossil-fueled generation was extensively augmented by hydroenergy and nuclear fission.

The continued rapid growth of average per caput electricity consumption in low- and medium-income economies will be the only way to narrow the existing disparities. In 2015 the U.S. annual per caput average was more than 13 MWh, Japan's was more than 8 MWh, and Europe averaged about 7 MWh. In contrast, China's annual per caput mean was less than 4.5 MWh, India's just above one MWh, and in sub-Saharan Africa (excepting South Africa) it remains well below one MWh. Despite decades of electrification programs, more than 1.2 billion people (mainly in India, Southeast Asia, and sub-Saharan Africa) still do not have access to electricity. Consequently, the global disparity in average per caput electricity use is greater than in the case of primary energy consumption, and the need for future production increases is more acute. This is perhaps most vividly portrayed by composite night-time satellite images, which starkly contrast brightly-lit affluent countries with the huge areas of darkness, or at best sparse light, over large parts of Asia, Africa, and Latin America.

While the overall efficiency of energy use in low- and medium-income countries is dismally low and should be greatly improved through technical innovation and better management, future higher energy needs cannot be met solely, or even mostly, through higher efficiency. Positive steps in this direction are essential. China's post-1980 achievements, more than halving the energy use per unit of Gross Domestic Product (GDP) show their fundamental importance: without them, China would be now consuming more than twice as much energy for every unit of its economic product as it does, as well as burdening its already highly degraded environment with even more pollutants. High-efficiency conversions clearly benefit economies and the environment, but they reduce overall energy use only on an individual or household level, or for a single company, particular industrial process, or entire production sector.

On national and global levels, the record shows the very opposite; there is no doubt that higher efficiencies of energy conversions have led to steadily greater consumption of fuels and electricity. This paradox was noted for the first time by Stanley Jevons (1835–1882), a prominent English economist, in 1865. In his words: "It is wholly a confusion of ideas to suppose that the economical use of fuels is equivalent to a diminished consumption. The very contrary is the truth." Jevons illustrated the phenomenon by contrasting the huge efficiency improvements of eighteenth-century steam engines (from Savery's and Newcomen's extremely wasteful machines to Watt's improved design) with the large increases in British coal consumption during the same period.

Two examples illustrate this common phenomenon for modern energy-consuming activities. First, in the year 2000, the average American passenger vehicle (including SUVs) consumed nearly forty percent less fuel per kilometer than in 1960, but more widespread ownership of automobiles and the higher annual average distance driven (roughly 17,000 km, compared to 15,000 km in 1960) resulted in an increase of average per caput consumption. Second, during the twentieth century, the efficiency of British public street lighting rose about twenty-fold, but the intensity of this illumination (MWh per km of road) rose about twenty-five times, more than eliminating all efficiency gains. As a result, average per caput energy consumption continued to rise, albeit at a slower pace (as expected), even in mature, post-industrial economies.

During the 1990s, despite deep economic problems and the stagnation of its GDP, Japan's average per caput energy consumption grew by fifteen percent; in the same period the already extraordinarily high U.S. and Canadian rates grew by about 2.5 percent, and France's by nearly ten percent. But things have changed since the year 2000, and in 2015 the average per caput energy consumption in all major Western economies (U.S., Canada, Germany, France) and in Japan was either slightly below, unchanged or only marginally above the 1990 rates: affluent economies might be, finally, reaching their energy consumption plateaux. And affluent countries could reduce their energy use by a third and still remain well-off. That is not the case with lower income economies: since 1990 per caput energy consumption has soared in China and it has grown, at slower rates, in India as well as in the relatively best performing economies in Africa – and without further substantial gains those nations could never modernize. But they now face the entirely new constraint of rapid global warming, something that modernizing Europe, North America and Japan did not have to worry about in the past.

We have three choices if we wish to keep on increasing global energy consumption while minimizing the risks of anthropogenic climate change due to rising combustion of fossil fuels and keeping atmospheric levels of greenhouse gases from rising to as much as three times their pre-industrial level: we can continue burning fossil fuels but deploy effective methods of sequestering the generated CO_2, we can revive the nuclear option, or we can turn increasingly to renewable energy. None of these options is yet ready to take over from fossil fuels on requisite scales, none could be the sole near-term solution, and all have their share of economic, social, and environmental problems.

Despite a great deal of theoretical research, and much interest shown by industries and governments, CO_2 sequestration is only in early deployment stages. In 2015, fifteen sequestration projects were in operation and more than twenty under construction or in planning stages: if all of them were working by 2020 they would be removing an equivalent of a mere 0.2 percent of 2015 emissions. Even if that were followed by rapid construction of new projects, the near-term effect will be marginal and the eventual contribution of these techniques to the management of the global warming challenge remains uncertain. In contrast, we have more than half a century of experience of large-scale commercial generation of nuclear electricity, which has shown us what to avoid and what techniques to favor. The general expert consensus is that any development of the nuclear industry cannot be a replica of the first generation; there has been no shortage of new, ingenious designs aimed at minimizing or eliminating the concerns that contributed to the stagnation (and in some countries even retreat) of nuclear electricity generation. Several, so-called, inherently safe nuclear reactor designs provide passive guarantees of fail-proof performance: even operator error could not (as it did in Chernobyl) lead to a core meltdown. Large-scale adoption of nuclear reactors would be made easier by flexible sizing that would make it possible to deploy small modular units. But the future of the industry will not depend primarily either on better designs (they have been available since the mid-1980s), or on the fears of catastrophic accidents (risk that must be compared with hundreds of safely operating reactors and with risks of other energy conversions). What has to change is the public acceptance of this potentially risky but very rewarding form of electricity generation, and I have argued that there is little chance of any substantial worldwide return to nuclear generation unless it will be led by the world's largest economy. But in 2017, U.S. nuclear plans seem no less confused and uncertain than they were ten or twenty years ago: there is no true strategic planning, and no sign that the public distrust of nuclear generation has eased. As the endless wrangling about the location and operation of the country's permanent repository of high-level radioactive wastes shows, the combination of executive intents, legislative delays, and legal appeals makes for decades of irresolution and offers little hope for any determined state-sanctioned nuclear-based solution to the country's future electricity needs.

All that may change, but not because the public finally appreciates the real relative risks of various electricity-generating options, as these have been known for decades, but because of the need for an accelerated decarbonization of the global energy supply required to deal with exceptionally rapid global warming. The nuclear option is not greenhouse gas free: we need coke to make the plant's many steel components, and the cement for its massive concrete structures comes from fossil fuel-fired kilns. But in comparison with today's dominant (coal-fired) mode of generation, nuclear plants produce at least ninety-five percent less CO_2 per unit of electricity. If our civilization were to face a true global warming shock, this would be very appealing. Consequently, the most rational strategy of future energy supply would be to combine improvements in conversion efficiency (particularly in industrialized economies) with reduced rates of overall energy demand (especially in affluent countries), keep the nuclear option open during the development of innovative reactor prototypes, and increase the contributions of non-fossil sources as quickly as economically feasible and environmentally acceptable. Because capital investment considerations and infra-structural inertia mean that it takes several decades for any new energy source or conversion to claim a substantial share of the market, we should not waste any time in aggressively developing and commercializing suitable renewable options.

Renewable energies: biomass, water, wind, solar

Biomass energies could only become an important component of future energy supply after the development of large-scale, intensive production of selected crop and tree species convertible, by advanced techniques, into liquid or gaseous fuels or electricity. This strategy has three fundamental drawbacks. First, as explained in Chapter 2, photosynthesis operates with an inherently very low power density, and hence any large-scale biomass fuel production would claim extensive areas of farmland (and it would have to be farmland, rather than marginal land, to sustain high productivity). Second, humanity already claims a very high share (most likely close to one-fifth) of the biosphere's net primary productivity (through harvests of food, feed, wood, grazing, and deliberately set grassland, and forest fires), and adding a further burden through massive fuel production would lead to a further loss of biodiversity and greater environmental degradation. Finally, the overall costs (economic, energetic, and environmental) of large-scale biomass energy production are very high.

Low power densities translate into very large land require-ments (Figure 30). The U.S. has been diverting about forty percent of corn, its largest crop, to ethanol fermentation but the produced fuel is an equivalent of less than ten percent of the country's gasoline consumption. Moreover, very few countries have enough farmland to even contemplate such mass-scale biofuel production. For example, replacing just a quarter of the world's 2015 fossil fuel consumption with cultivated woody biomass would require (even with high average yields of 15 t/ha) tree plantations of more than 500 million hectares, roughly equal to the total combined forested land in Europe, the U.S. and Canada, clearly an impossible option. Devoting even limited areas to biomass crops would be irrational for the scores of densely populated countries that already have shortages of the arable land needed to secure their basic food supply and so are major

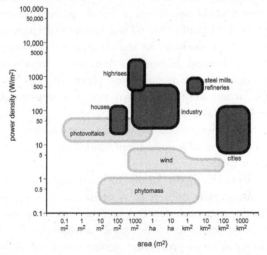

Figure 30 Comparison of power densities of energy consumption and renewable energy production

food importers. Creating new biomass plantations would lead to further loss of natural grasslands, wetlands, and the lowland tropical forests.

Moreover, modern liquid biofuels (required to displaced fuels refined from crude oil) have very low EROI (ranging from less than 1.5 to less than five) due to the combined energy cost of machinery, fertilizers, irrigation, and biomass conversion. But even the cultivation of those biomass crops that produce relatively high net energy gains would still have undesirable environmental impacts, above all increased soil erosion, soil compaction, and contamination of aquifers and surface waters by nitrogen and phosphorus lost from fertilizers, causing aquatic eutrophication (that is, the enhanced growth of algae, which disrupt the existing ecosystem).

Conversion of waste cellulosic biomass (logging and lumber mill residues, and crop residues not needed for protecting soils against excessive erosion and recycling nutrients) is the best choice, but scaling up this (just started) production to displace a significant fraction (say fifteen to twenty percent) of currently used fuels derived from crude oil will take time. Expanding fuel-wood groves for household use and planting fast-growing species for commercial wood deliveries is desirable in areas with good growing conditions, or regions with plenty of available barren slope land, where afforestation may not only improve regional fuel supply but also reduce soil erosion. But any dreams of modern megacities fuelled by woody biomass should remain, for the sake of a reliable food supply and limited environmental impacts, just that.

Hydrolectricity is the largest modern non-fossil source of primary energy; the combination of relatively low cost, high suitability to cover peak demand, and the multi-purpose nature of most large reservoirs (they serve as sources of irrigation and drinking water, a protection against downstream flooding, recreation sites, and, increasingly, places for aquacultural production) should make it one of the most desirable choices in a world moving away from fossil fuels. This conclusion seems to be strengthened by the fact that on the global scale most of this clean renewable energy resource remains untapped: the International Commission on Large Dams put the global potential of economically feasible projects at just over 8 PWh, roughly twice the current rate of annual generation. As expected, the remaining potential is unevenly distributed. Europe, North America, Australia, and Japan have already developed as much of their large-scale hydrogenerating capacity as they can (there is always the potential for microstations), but Latin America has so far tapped less than a quarter, Asia less than a fifth, and Africa not even a twentieth, of their respective potentials.

This untapped potential would seem especially welcome, as it is precisely those continents where future demand will be highest, but it now appears that the development of hydrogeneration in those regions will not proceed either as rapidly or as exhaustively as was assumed during the closing decades of the twentieth century. In an important shift of perception, hydroenergy has changed from a clean, renewable and environmentally benign resource to a much more controversial cause of socially and environmentally disruptive, and economically questionable, developments. As a result, international and internal opposition to megaprojects (plants with multigigawatt capacities) has spread, and the willingness of governments and international lending agencies to finance such developments has declined. Sweden has banned further hydrostations on most of its rivers, Norway has set aside all existing plans, in the U.S., since 1998, the decommissioning rate for large dams has overtaken the construction rate, and many countries in Asia (most notably in India) and Latin America have seen vigorous public protests against new projects.

CONCERNS ABOUT LARGE DAMS

In 2000, the World Commission on Dams published a report which stressed that all future projects should consider social and environmental effects to be as important as the, traditionally dominant, economic benefits of electricity generation (or of irrigation or water supply). While some recent criticism has been ideologically motivated and clearly overwrought, there is no doubt that large hydroprojects bring a number of serious social and environmental changes. In Chapter 4 I noted the major concerns: the large numbers of people displaced by the creation of major reservoirs, the excessive silting of many storages, the aging of average river runoff and the fact that water reservoirs are (much like fossil fuels) sources of greenhouse gas emissions, as they release CO_2 and CH_4 from submerged and decaying trees and shrubs.

A new concern has emerged, as we see more indications of the inevitable deterioration of aging dams and contemplate the costs of their eventual decommissioning: these matters were given no, or insufficient, attention when they were built. We can only speculate about the ultimate life expectation of such massive structures, and have no good strategies to deal with the excessive silting and premature filling of reservoirs, which reduces their useful life span (in parts of monsoonal Asia affected by severe deforestation the process has already cut the expected duration of some reservoirs by as much as half). All this makes it much more unlikely that the remaining hydrogeneration potential will be developed as aggressively as it was in the twentieth century.

But even without any obstacles to their construction, new hydrogeneration capacities could supply only a part of the expected demand, and then only by claiming large expanses of river valleys, forests, grasslands, settlements, and agricultural land. The average power density of existing hydrostations (actual generation rather than installed capacity: this adjustment is necessary because dry years curtail generation at many dams) equates to less than 3 W/m² and they claim nearly 150,000 km² of land. If all of the remaining potential were to be realized during the first half of the twenty-first century, new reservoirs would claim roughly 500,000 km², an area as large as Spain. But hydroenergy can be also harnessed on a smaller scale, and many Asian, African, and Latin American countries have excellent potential for developing stations, with capacities less than 10–15 MW, which would not make much of a dent in a nationwide supply of a populous country, but could suffice to electrify a remote region or an island.

Wind energy, harnessed by large and efficient turbines, has emerged in the 1990s (less than a decade after a failed mini-boom during the 1980s) as the leading renewable energy choice, thanks largely to aggressive promotion and adoption in a number of Western European countries. Better designs, and larger sizes, of wind machines made a big difference: ratings rose from 40–50 kW during the early 1980s to 500–750 kW by the late 1990s, when the first turbines with capacities of more than 1MW went on line and by 2015 machines of 3-4 MW were common, with the largest offshore units at 8 MW. Danish designs (Vestas) have led the way, and the country also leads in per caput wind capacity. Worldwide totals of installed wind-generating capacity rose from 1 GW in 1985 to 432 GW in 2015, with China, U.S. and Germany in the lead. Average capacity factors have improved with better turbine designs to nearly twenty-five percent worldwide and to just over thirty percent in the U.S.. As already noted, in 2015 wind turbines generated just 3.5 percent of the world's electricity, but exponential growth of new capacities is expected to continue.

To avoid conflict, much of future wind power will be in large offshore wind farms (a number of them already operate in Denmark, Sweden, the Netherlands, and the U.K.), or it will come from re-powering old sites with larger turbines. The ultimate extent will depend on the progress of integrating intermittent electricity sources into national and continental grids through high-voltage interconnections and large-scale electricity storage. Environmental concerns associated with large turbines range from the well-documented risks to migrating birds, to esthetic objections, both to turbines massed in large onshore wind farms and the size of the largest machines. Well-designed and well-maintained turbines should work for more than two decades but we have to accumulate operating experience with a very large number of offshore units to be able accurately to assess the long-term availability and reliability of these machines exposed to such hazards as hurricanes, heavy icing and air laden with salt aerosols.

Compared to wind-powered electricity, photovoltaics is still a much smaller contributor, accounting for just one percent of all electricity generation in 2015 – but the installations of new PV modules have been expanding faster than those of new wind turbines. China, U.S., Germany, Japan and Italy are the leaders but the power ratings of PV units are not directly comparable with other modes of electricity generation, because they are expressed in peak watts measured under high irradiance ($1,000 \text{ W/m}^2$, the equivalent of mid-day, clear-sky insolation) rather as an average performance.

Three fundamental reasons make the PV conversion of solar radiation into electricity the most appealing of all renewable sources: the unparalleled magnitude of the resource, its relatively high power density, and the inherent advantages of the conversion technique (no moving parts, silent operation at ambient temperature and pressure, and easy modularity of units), but the two key drawbacks are its still relatively low conversion efficiency and low capacity factors in less sunny regions. Efficiencies of PV cells have risen from less than five percent during the early 1960s, when the first modules were deployed on satellites, to twenty-five percent for high-purity silicon crystals in the laboratory, but the field efficiencies are around fifteen percent. PV films, made of amorphous silicon (or gallium arsenide, cadmium telluride, or copper indium diselenide), have reached as much as twenty-two percent in the laboratory, but deliver eleven to thirteen percent in field applications. Declining costs of PV cells have made them particularly competitive in sunny locations where their capacity factor can average twenty five percent (compared to just over ten percent in Germany).

More efficient photovoltaic cells would be most welcome because of their relatively high power densities: efficiencies close to twenty percent would translate to electricity generation rates between 20–40 W/m^2, two orders of magnitude better than biomass conversion, and one better than most hydro and wind projects. The future rapid growth of PV installations is assured but, as in the case of wind turbines, their eventual share of electricity generation will depend on their successful integration of this intermittent energy source in large grids, and hence on the development of new high voltage interconnections and new (slowly improving) storage capabilities.

Innovations and inventions: impossible forecasts

The most welcome advance would be a large-scale affordable means of electricity storage: with it a combination of wind-driven and PV electricity generation could provide a significant share of electricity supply. Capacities of large storage batteries have gone up but by 2015 the best ratings were in tens of MW (of massed lithium-ion batteries) dischargeable over several hours. What is needed, particularly for the world's growing megacities, are storages on GW scale — and pumped storage still remains the only effective way of doing that. This uses two water reservoirs at least several hundred meters apart in height; electricity not needed by the grid is used to pump water from the lower to the upper storage, where it is kept until released for generation during periods of peak demand. The worldwide capacity of pumped storage is close to 130 GW; the largest unit (Bath County in Virginia) rates 3 GW. But pumped storages are expensive, and the requirement for reservoirs in high relative elevations makes them inconceivable in densely populated lowlands. There is another option that would obviate electricity storage: inexpensive intermittently generated electricity could be used to electrolyze water and variable flows of renewable energies would produce hydrogen as a major energy carrier.

HYDROGEN AS ENERGY CARRIER

Hydrogen cannot, contrary to what so many popular writings repeatedly imply, be a significant source of energy. Unlike methane, it is not present in huge reservoirs in the Earth's crust, and energy is needed to produce it, from either methane or water. But some of its properties make it an outstanding energy carrier. Its key advantages are superior energy density (liquid hydrogen contains 120 MJ/kg compared to 44.5 MJ/kg for gasoline), a combustion that yields only water, and the possibility of using it in fuel cells.

The key advantages of fuels cells (electrochemical devices that combine hydrogen and oxygen to produce electricity) are the absence of moving parts, a quiet and highly efficient (commonly in excess of sixty percent) operation, and their modularity (they can be made small enough to power a laptop or large enough to generate electricity in multi-megawatt plants). An enormous amount of research interest in fuel cells has led to exaggerated expectations of their early commercialization, but their cost (except for a few relatively small niche markets) is still high, only a few models are available and refueling infrastructure is extremely sparse — and unless it will be in place, carmakers will be reluctant to mass-produce hydrogen-powered fuel cell cars. The transition to hydrogen-powered vehicles would also be complicated by the need for energy-dense storage and safe handling. Uncompressed hydrogen occupies 11,250 l/kg; pressurizing it into a high pressure steel tank reduces this to 56 l/kg, but this is equivalent to less than three liters of gasoline, or enough fuel to move an efficient compact car less than fifty kilometers. Liquefied hydrogen occupies only 14.1 l/kg but needs to be kept below −241°C, an immense engineering challenge in a small vehicle. Adsorption on special solids with large surface areas, or absorption by metal hydrides seem to be the most promising options.

The safety of hydrogen distribution is no smaller challenge. While the highly buoyant gas leaks quickly and it is non-toxic (making its spills more tolerable than those of gasoline) its ignition energy is only one-tenth that of gasoline, its limit of flammability is lower, and its range of flammability much higher. These will mean much stricter precautions at hydrogen stations than those now in place at gasoline filling stations.

Moving toward a system dominated by hydrogen is clearly consistent with the long-term decarbonization of the modern energy supply, but the progress will be gradual and we should not expect any large-scale transition to a hydrogen economy during the coming generation (Figure 31). The hydrogen:carbon (H:C) ratio of dominant fuels has moved from around 0.5 in

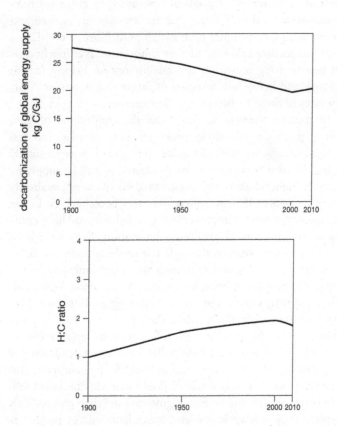

Figure 31 Decarbonization of the world's energy supply (calculated from data in UN and BP statistics)

wood (for actually oxidized H atoms), to about 1.0 in coal, and 1.8 for liquid hydrocarbons. The continuation of this trend points first to the emergence of natural gas (with H:C of 4) as the leading fossil fuel, and eventually (but almost certainly not during the first half of the twenty-first century) perhaps to a hydrogen-dominated world. But trends can be derailed or accelerated by social or political upheavals, or enter frustrating *culs-de-sac*, and only those that are strongly entrenched and rely on mature techniques have a high probability of continued adoption, accompanied by further innovation. Neither hydrogen nor a strong revival of the nuclear option belong to this category, and hence any forecasts of future milestones or diffusion rates of these techniques are just guesses.

In contrast, there is no doubt that the combustion of fossil fuels – gradually becoming more efficient, cleaner, and less carbon-intensive – will dominate the global energy supply during the next two generations. As electricity will be supplying a steadily higher share of the world's final use of energy, its already generally highly efficient conversions will become even better. The greatest room for improvement is in lighting, and light emitting diodes are a most promising innovation. They have been around for many years as the little red or green indicator lights in electronic devices, and (although you may think you have a light bulb there) then became common in car brake lights, taillights, and turn signals, and also in traffic lights. But since 2012 they have started, finally, to make their greatest impact as long-lasting, highly efficient (and also full-spectrum) replacements of conventional incandescent light bulbs. So our grandchildren will use lights that may be, on average, at least fifty percent more efficient than ours. There is also little doubt that our continued reliance on fossil fuels will be first augmented and then progressively supplanted by renewable energies: major hydroenergy projects in Asia and Africa and by wind-powered electricity generation and PV conversions on all continents (Figure 32).

Figure 32 Brazos wind farm in West Texas with 160 one-megawatt turbines (reproduced from Wikipedia)

Inexpensive intermittently generated electricity integrated into large grids and relying on considerable storage capacity can take over large shares of markets now served by fossil fuels (including, eventually, passenger cars) but there are no near-term prospects that it could displace high energy density fossil fuels in such major sectors as intercontinental flight and shipping (both for container and bulk cargo vessels). And we have no readily deployable commercial techniques to produce such fundamental materials as pig iron, cement or ammonia on required scales without fossil fuels. But history makes it clear that the train of human ingenuity is not about to stop. Although major inventions tend to come in irregularly spaced clusters rather than an orderly progression, half a century is long enough to see the emergence, and even substantial diffusion, of several new inventions whose universal adoption could transform the energy foundations of late twenty-first-century civilization. Such developments are highly probable, but their nature and their timing are entirely unpredictable. Remember the two major late twentieth-century examples: neither the emergence of mass air travel (thanks to the invention of the gas turbine and its much improved turbofan designs) nor the invention of solid state electronic devices (transistors, integrated circuits, and microprocessors) were predicted to bring major technical and economic advances and social changes even just a few years before their commercial debuts!

The transition to an energy system based predominantly on non-fossil resources is in only its earliest phase. In some ways this appears to be a greater technical and social challenge than the last epochal shift (from animate energies and biomass fuels to coal, hydrocarbons, engines, and electricity). Given the knowledge and resources at our command, this challenge should be manageable but the shift will not be (it cannot be) extraordinarily rapid. We now have much more powerful scientific and technical means to come up with new solutions, and we also have the benefits of unprecedented information sharing and international co-operation, and can take advantage of various administrative, economic, and legal tools aimed at promoting the necessary adjustments, from more realistic pricing to the sensible subsidies required to accelerate the diffusion new and promising techniques or help them to achieve a critical market mass more quickly. But the task ahead is daunting, because the expectations for energy futures are high. They combine the anticipation of continued supply improvements (in access, reliability, and affordability) in already affluent (or at least fairly well-off) countries (whose populations total about one billion), not only with the necessity of substantial increases in average per caput energy consumption among the world's (now more than five billion) less fortunate people, but also with the need to harness, distribute, and convert these massive energy flows in ways compatible with long-term maintenance (and in many cases major enhancement) of local and global environmental quality. Such challenging, fundamental transformations offer the best opportunities for creative solutions and effective adaptations. The evolutionary and historical evidence shows that humans are uniquely adapted to deal with change. While our past record of ingenuity, invention, and innovation is no guarantee that another fairly smooth epochal energy transition will take place during the next few generations – it is a good foundation for betting that our chances are far better than even.

Further readings and data sources

Given the range of topics that can be studied from the perspective of energy even a far from exhaustive bibliography limited to works published since 1990 could be longer than a beginner's guide. The following recommendations (eight for each of five topical sections), and a short list of leading websites containing data and statistics, cover all major concerns of recent energy studies.

Energy basics and overviews

Cleveland, C. and C. Morris. 2013. *Handbook of Energy* (in two volumes). Waltham, MA: Elsevier.

Fricke, J. and W.L. Borst. 2013. *Essentials of Energy Technology*. Weinheim: Wiley-VCH.

Goldemberg, J. 2012. *Energy: What Everyone Needs to Know*. Oxford: Oxford University Press.

Niele, F. 2005. *Energy: Engine of Evolution*. Amsterdam: Elsevier.

Schobert, H. 2013. *Energy: The Basics*. London: Routledge.

Smil, V. 2000. *Energies: An Illustrated Guide to the Biosphere and Civilization*. Cambridge, MA: MIT Press.

Smil, V. 2008. *Energy in Nature and Society: General Energetics of Complex Systems*. Cambridge, MA: MIT Press.

Smith, J.C. 2015. *Energy, the Subtle Concept*. Oxford: Oxford University Press.

Energy in life

Alexander, R.M. 1999. *Energy for Animal Life*. Oxford: Oxford University Press.

Biewener, A.A. 2003. *Animal Locomotion*. Oxford: Oxford University Press.

Butz, S. 2014. *Energy and Agriculture: Science, Environment, and Solutions*. Clifton Park, NY: Delmar Publishers.

Food and Agriculture Organization. 2004. *Human Energy Requirements*. Rome: FAO.

Gibney, M.J. et al. 2009. *Introduction to Human Nutrition*. Hoboken, NJ: Wiley-Blackwell.

Nicholls, D.G. and S. Ferguson. 2013. *Bioenergetics*. Cambridge, MA: Academic Press.

Pielou, E.C. 2007. *The Energy of Nature*. Chicago: University of Chicago Press.

World Health Organization. 2002. *Protein and Amino Acid Requirements in Human Nutrition*. Geneva: WHO.

Energy in history

Kander, A. Malanima, P. and P. Warde. 2013. *Power to the People: Energy in Europe Over the Last Five Centuries*. Princeton, NJ: Princeton University Press

Oleson, J.P., ed. 2008. *The Oxford Handbook of Engineering and Technology in the Classical World*. Oxford: Oxford University Press.

Rosen, W. 2012. *The Most Powerful Idea in the World: The Story of Steam, Industry, and Invention*. Chicago: University of Chicago Press.

Smil, V. 2005. *Creating the Twentieth Century: Technical Innovations of 1867–1914 and Their Lasting Impact*. New York: Oxford University Press.

Sørensen, B. 2011. *History of Energy: Northern Europe from the Stone Age to the Present Day*. London: Routledge.

Weissenbacher, M. 2009. *Sources of Power: How Energy Forges Human History*. Santa Barbara, CA: Praeger.

White, P. and T. Denham, eds. 2006. *The Emergence of Agriculture: A Global View*. London: Routledge.

Wrangham, R. 2009. *Catching Fire: How Cooking Made Us Human*. New York: Basic Books.

Energy in modern world

Bret, A. 2014. *The Energy-Climate Continuum*. New York: Springer

Hausman, W.J., Hertner, P. and M. Wilkins. 2008. *Global Electrification*. Cambridge: Cambridge University Press.

Jones, C.F. 2014. *Routes of Power: Energy and Modern America*. Cambridge, MA: Harvard University Press.

Lovins, A. and Rocky Mountain Institute. 2011. *Reinventing Fire*. White River Junction, VT: Chelsea Green Publishing.

Smil, V. 2006. *Transforming the Twentieth Century: Technical Innovations and Their Consequences*. New York: Oxford University Press.

Smil, V. 2010. *Prime Movers of Globalization: The History and Impact of Diesel Engines and Gas Turbines*. Cambridge: MIT Press.

Smil, V. 2015. *Power Density: A Key to Understanding Energy Sources and Uses*. Cambridge, MA: MIT Press.

Yergin, D. 2008. *The Prize: The Epic Quest for Oil, Money, and Power*. New York: Simon and Schuster.

Energy information and statistics

British Petroleum. *Statistical Review of World Energy*. www.bp.com/en/global/corporate/energy-economics/statistical-review-of-world-energy.html

International Energy Agency. *Statistics*. www.iea.org/statistics/

International Renewable Energy Agency. http://www.irena.org

Organization of Petroleum Exporting Countries. http://www.opec.org
United Nations. *Energy Statistics*. http://unstats.un.org/unsd/energy
Unites States Energy Information Administration. www.eia.gov
World Bank. *Energy*. www.worldbank.org/en/topic/energy
World Coal Association. www.worldcoal.org/
World Energy Council. www.worldenergy.org/
World Nuclear Association. http://world-nuclear.org/

Index